Advances in Heterocatalysis by Nanomaterials

Advances in Heterocatalysis by Nanomaterials

Special Issue Editors

Ioannis V. Yentekakis

Wei(Willy) Chu

MDPI • Basel • Beijing • Wuhan • Barcelona • Belgrade • Manchester • Tokyo • Cluj • Tianjin

Special Issue Editors
Ioannis V. Yentekakis
Technical University of Crete
Greece

Wei(Willy) Chu
Sichuan University
China

Editorial Office
MDPI
St. Alban-Anlage 66
4052 Basel, Switzerland

This is a reprint of articles from the Special Issue published online in the open access journal *Nanomaterials* (ISSN 2079-4991) (available at: https://www.mdpi.com/journal/nanomaterials/special_issues/nano_heterocatalysis).

For citation purposes, cite each article independently as indicated on the article page online and as indicated below:

LastName, A.A.; LastName, B.B.; LastName, C.C. Article Title. *Journal Name* **Year**, *Article Number*, Page Range.

ISBN 978-3-03928-835-9 (Pbk)
ISBN 978-3-03928-836-6 (PDF)

Contents

About the Special Issue Editors

Ioannis V. Yentekakis is currently Full Professor of Physical Chemistry and Catalysis in the School of Environmental Engineering, Technical University of Crete (TUC), Greece. He received Chemical Engineering Diploma (1983) and Ph.D. (1988) from the Dept of Chemical Engineering, University of Patras (UP). His academic career is also related to Princeton University NJ USA (Post Doc), ICE-HT/FORTH (Post Doc-Senior Researcher), Dept of Chemical Engineering UP (Assistant Professor), and Cambridge University UK (Visiting Professor). His current interests include the development of novel nano-materials and processes in heterogeneous catalysis for green chemistry and environmental protection, renewable energy generation, electrocatalysis and fuel cells, hydrogen energy, and natural gas/biogas valorization. He has authored >110 journal publications (>4500 citations, h-index=38; source: Google Scholar), three international patents, and 10 books, and has Guest Edited five Special Issues.

Wei(Willy) Chu is currently distinguished Full Professor in Sichuan University Dept of Chemical Engineering, China, and Senior Researcher (PI) at SABIC Tech. Center in Riyadh, Saudi Arabia. He received his Bachelor Degree from Nanjing University; his MSc from University of Lyon 1, France; and his Ph.D. from University of Strasbourg 1 (ULP), France. His current interests include applied and environmental catalysis, energy materials, nano functional materials, Fischer Tropsch synthesis and clean energy, and porphyrins and fine chemicals. He has authored >230 journal publications, which have received >9000 citations, h-index=48; source: Google Scholar.

nanomaterials

Editorial

Advances in Heterocatalysis by Nanomaterials

Ioannis V. Yentekakis [1,*] and Wei Chu [2]

1 Physical Chemistry & Chemical Processes Laboratory, School of Environmental Engineering, Technical University of Crete (TUC), 73100-Chania, Crete, Greece
2 Key Laboratory of Green Chemistry and Technology of Ministry of Education (MOE), College of Chemical Engineering, Sichuan University, Sichuan 610065, China; chuwei1965@scu.edu.cn
* Correspondence: yyentek@isc.tuc.gr; Tel.: +30-28210-37752

Received: 10 March 2020; Accepted: 24 March 2020; Published: 26 March 2020

Overview

Heterogeneous catalysis played, plays, and will continue to play a major key role in industrial processes for the large-scale synthesis of commodity chemicals of global importance [1,2] and in catalytic systems that possess a critical role in energy generation [3–8] and environmental protection approaches [9–13]. As a result of the ongoing progress in materials science, nanotechnology, and characterization methods, great advances have been currently recorded in heterogeneous catalysis by nanomaterials, the so-called "nanocatalysis" [14–16]. Efficient approaches and advanced methods for the design of nanostructured composite materials (up to atomic level) [14–18], subject to specific nanomorphologies with enhanced metal–metal and metal–support interactions that are favorable for catalysis (that enable fine-tuning of the critical physicochemical properties of the designed catalysts) [18], provide optimized catalytic systems with outstanding performances in numerous eco-friendly and cost-effective applications. Under this line, great progress has been achieved in many applications of heterogeneous catalysis involving, for example, emissions control catalysis, waste treatment, photocatalytic, biorefinery, CO_2 utilization, and fuel cells applications, as well as hydrocarbon processing for H_2, added-value chemicals, and liquid fuels production, among several others [1–18].

This Special Issue: Contributions and Highlights

In this context, this Special Issue of Nanomaterials has succeeded in collecting 10 high-quality contributions that cover recent research progress in the field, involving rational synthesis and adequate characterization of novel nanostructured catalysts with improved efficiency and performance in several high-impact environmental, energy, added-value chemicals/organics synthesis and biotransformation processes, including: (i) the synthesis of added-value chemicals/organics and biotransformations (5 papers), (ii) photocatalytic pollutants degradation (2 papers), (iii) photocatalytic or electrocatalytic water splitting for H_2 and/or O_2 evolution (2 papers), and (iv) wastewater cleaning from pharmaceuticals (1 paper). The apparent dispersion of the application subjects and targets of these 10 paper contributions declares the prospect and importance of nanomaterials in all the directions of the area of heterogeneous catalysis.

(i) Nanomaterials for Chemicals/Organics Synthesis and Biotransformation Applications

Su et al. [19] synthesized Rh nanoparticle catalysts dispersed on TiO_2 and boron-decorated TiO_2 nanotube supports (Rh/TNTs, Rh/B-TNTs), which were evaluated on the hydroformylation of 2-methyl-3-butennitrile (2M3BN). Given that 2M3BN can be used for the preparation of adiponitrile after isomerization, and the hydroformylation of 2M3BN is an industrially important and scientific research subject, in addition to the fact that hydroformylation processes are related with difficulties upon recovery of the homogeneous catalysts employed in industrial practice, the present study, employing heterogeneous catalysts, is of considerable interest. A variety of techniques, namely X-ray diffraction (XRD), scanning electron microscopy (SEM), transmission electron microscopy (TEM), X-ray photoelectron spectroscopy (XPS), temperature-programmed desorption of ammonia (NH_3-TPD), inductively coupled plasma–atomic emission spectroscopy (ICP-AES), and Brunauer–Emmett–Teller (BET) were used for the characterization of catalysts. Superior olefin hydroformylation catalytic performance (activity, selectivity to aldehydes) of Rh/B–TNTs in comparison to Rh/TNTs was obtained, resulting from its improved acidity. A better dispersion of Rh nanoparticles on Rh/B-TNTs than that on Rh/TNTs was achieved with average particle sizes of 2.8 and 4.9 nm, respectively, while B-modification of the support also improved catalytic stability.

Riyadh et al. [20] synthesized chitosan (a natural polysaccharide)–MgO hybrid nanocomposite by a simple, one-pot precipitation method, which was characterized by means of Fourier transform pectroscopy (FTIR), elemental analysis (EDX), and scanning electron microscopy (SEM). This chitosan nanocomposite is a three-dimensional, cross-linked, polymeric matrix of chitosan with active NH and OH functional groups, incorporating MgO nanoparticles, and it can be used as a novel basic heterogeneous catalyst in the form of a solid film. The material served here as a powerful eco-friendly basic catalyst under microwave irradiation in the synthesis of two novel series of 5-arylazo-2-hydrazonothiazoles 4a–j and 2-hydrazono [1,3,4] thiadiazoles 8a–d, incorporating a sulfonamide group. Its catalytic performance, as a green recyclable catalyst, was comparatively evaluated by means of triethylamine (a traditional catalyst). The significantly better yields of the chitosan-MgO toward hydrazonothiazoles and hydrazono [1,3,4] thiadiazoles was attributed to the obtained nanosized MgO and the synergistic effect that is created by the combination of the basic nature of both MgO and chitosan. The novel nanocomposite catalyst can be easily recovered and reused for many times without loss in its catalytic activity, therefore making it promising for implementation in many other organic transformations.

Zhao et al. [21] produced ZrO_2 nanoparticles, ZrO_2(P) and ZrO_2(H), with different tetragonal phase content (higher in the former), which was used as support for the preparation of 10 wt % Ni/ZrO_2 catalysts via impregnation. The catalysts were characterized by means of XRD, Raman, H_2-TPR, XPS, and H_2-TPD techniques, and their catalytic performance was evaluated under the hydrogenation of maleic anhydride toward succinic anhydride and γ-butyrolacetone. The Ni/ZrO_2(P) catalyst exhibited stronger metal–support interactions than the Ni/ZrO_2(H) due to its higher number of oxygen vacancies and the low-coordinated oxygen ions on its surface, resulting in smaller Ni particles and higher C=C hydrogenation activity for maleic anhydride to succinic anhydride. However, the C=O hydrogenation activity of Ni/ZrO_2(P) was found to be much lower than that of the Ni/ZrO_2(H). A 43.5% γ-butyrolacetone yield over the Ni/ZrO_2(H) versus a much lower of only 2.8% over the Ni/ZrO_2(P) catalyst, at 210 °C and 5 MPa of H_2 pressure, was obtained. In situ FTIR characterization revealed that the high C=O hydrogenation activity for the Ni/ZrO_2(H) could be attributed to the surface synergy between active metallic nickel species and relatively electron-deficient oxygen vacancies. The obtained insight could stimulate new strategies for ZrO_2-based catalysts performance optimization under α, β-unsaturated aldehyde, and ketone hydrogenation reactions by modulating the surface structure of ZrO_2 supports.

Wu et al. [22] investigated a series of zeolites with different topology structures, including SAPO-34, SUZ-4, ZSM-5, USY, MOR, and beta, as catalysts for the synthesis of polyoxymethylene dimethyl ethers (PODEn) from dimethoxymethane (DMM) and trioxymethylene (TOM). Both experimental and theoretical studies were employed to evaluate the influence of acidic properties and textural/morphological characteristics of the zeolites on their activity. It was confirmed that a pore mouth diameter larger than a TOM molecule was an essential prerequisite for the synthesis of PODEn over zeolites; the synergistic effect between medium-strong Brønsted acid sites (Brønsted MAS) and the maximal available space of zeolites determines the catalytic performance of all zeolites studied. Mechanistic implications involve first the decomposition of DMM and TOM into methoxymethoxy groups (MMZ) and CH_2O monomer over Brønsted MAS. Then, the steric constraint of the maximum included sphere, with an appropriate size in zeolite channels, can promote the combination of CH_2O and MMZ to form transition species $ZO(CH_2O)_nCH_3$, which reacted with the methyl-end group to form PODEn over Brønsted MAS. The reaction temperature appeared to affect product distribution (selectivity) due to changes in the activity of intermediate species, which also strongly depends on the maximum available space in zeolite channels. Overall, it is concluded that a pore mouth diameter larger than the TOM molecule, a proper amount of Brønsted MAS, and an appropriate maximum including the sphere size are necessary conditions to obtain high PODEn selectivity at low temperatures.

Gournis, Stamatis, and co-workers [23] synthesized porous carbon cuboids (PCC) and functionalized PCCox in order to develop novel β-glucosidase-based nanobiocatalysts for the bioconversion of oleuropein to hydroxytyrosol. Using non-covalent or covalent immobilization approaches, β-glucosidases from almonds and thermotoga maritima were attached for the first time on oxidized and non-oxidized porous carbon cuboids (PCC). A variety of characterization methods including Fourier transform infrared spectroscopy (FTIR), X-ray photoelectron spectroscopy (XPS), and atomic force microscopy (AFM) were employed for the adequate characterization of the bionanoconjugates. The oxidation state of the oxidized PCCs as a type of nano-support and the immobilization procedure appeared to play a key role on the immobilization efficiency or the catalytic activity of the immobilized β-glucosidases. The novel bionanoconjugates nanobiocatalysts formed efficiently catalyzed the hydrolysis of oleuropein, leading to the formation of its bioactivderivative, hydroxytyrosol, which is a phenolic compound with numerous health benefits. The bionanoconjugates exhibited high thermal and operational stability, up to 240 h of repeated use, pointing out their great prospect in various biotransformations.

(ii) Nanomaterials for Photocatalytic Pollutants Degradation Applications

Long, Yan, and co-workers [24] in an attempt to meet the urgent need for advanced photocatalytic materials, fabricated novel visible light-driven heterostructured $g\text{-}C_3N_4/TiO_2$ (CNT) composites, using graphitic carbon nitride ($g\text{-}C_3N_4$) as precursor and fibrous TiO_2 via the electrospinning preparation method. The photocatalytic performance of CNT was evaluated under the rhodamine B degradation and was found to be superior to that of the commercial TiO_2 (P25®) and electrospun TiO_2 nanofibers. The specific CNT heterostructure and its enlarged specific surface area enhanced the photocatalytic performance, suppressing the recombination rate of photogenerated carriers while broadening the absorption range of light spectrum. Heterostructured CNTs with an appropriate proportion can rationally use visible light and significantly promote the photogenerated charges transferred at the contact interface between $g\text{-}C_3N_4$ and TiO_2, rendering them as prospective candidates in photocatalytic pollutants degradation processes, as the authors conclude.

Under the same purpose, Bresolin et al. [25], based on its ability to absorb visible light, synthesized a methylammonium lead iodide perovskite (MAIPb), which was evaluated as a visible-light photocatalyst for the degradation of two model pollutants, rhodamine B (RhB) and methylene blue (MB). An approximately 65% photodegradation of RhB was achieved after 180 min of treatment, while the efficiency was enhanced up to 100% by assisting the process with a small amount of H_2O_2. The visible-light activity of the MAIPb perovskitic-structure photocatalyst was attributed to its outstanding optoelectronic properties, i.e., its ability to absorb light as well as to enhance the separation of photogenerated carriers.

(iii) Nanomaterials for Photocatalytic or Electrocatalytic Water-Splitting Applications

Zeng and coworkers [26] under the view of using photocatalysis as a green technique to convert solar energy to chemical energy, specifically H_2 production from water splitting, prepared ZnO and ZnO-red phosphorus heterostructures (ZnO/RP), through a facile calcination method for the first time. The materials studied in respect to their photocatalytic activity for H_2 evolution through water splitting offered considerable efficiency and excellent photostability under AM1.5 light irradiation. The w/w ratio of RP in the ZnO/RP heterostructure was used for material optimization. It was found that

ZnO/PR heterostuctures exhibit up to 20.8-fold enhancement of H_2 production compared to bare ZnO, and moreover overcome the photocorrosion sensitivity of ZnO. This privileged water-splitting activity and stability of ZnO/PR was considered to result from the rapid transfer and effective separation of photogenerated electrons and holes between the heterointerface of ZnO and RP and the inhibited charge carrier recombination on the surface.

Wen et al. [27] recognized the fact that the oxygen evolution reaction (OER), in other words the water-splitting reaction, is a pivotal step for many sustainable energy technologies, and in order to overcome common unfavorable sluggish kinetics and high overpotentials during OER, synthesized a hybrid electrocatalyst (CoNi-ZIF-67@$Ti_3C_2T_x$), i.e., an MXene supported CoNi-ZIF-67 hybrid. This was achieved by the in situ growth of bimetallic CoNi-ZIF-67 rhombic dodecahedrons on the $Ti_3C_2T_x$ matrix via coprecipitation. They found that the inclusion of the MXene matrix produces smaller CoNi-ZIF-67 particles and increases the average oxidation of Co/Ni elements endowing the CoNi-ZIF-67@$Ti_3C_2T_x$ with a pronounced OER electrocatalytic performance, which is much better than the IrO_2 electocatalysts and the pure CoNi-ZIF-67. Therefore, this work shows new strategies for the development of efficient non-precious metal electrocatalysts for OER.

(iv) Nanomaterials for Adsorption-Based Wastewater Cleaning from Pharmaceuticals

Yan and co-workers [28] worked on a subject of growing environmental attention, that is the removal of pharmaceuticals from wastewater, in particular antibiotics, which are stable, difficult to degrade, and are able to generate antibiotic-resistant genes in microorganisms with concomitant adverse effects in ecosystems. In this context, the authors prepared magnetic *N*-doped porous carbon (MNPC) via the self-catalytic pyrolysis of bimetallic metal–organic frameworks (MOFs) and studied its efficiency on antibiotics adsorption. The as-produced material showed favorable features (e.g., high surface area and pore volume, good graphitization degree, rich N-doping, and magnetic properties), allowing it to be an endowed antibiotic adsorbent. This was experimentally revealed by testing the adsorption capacity of MNPC on norfloxacin (NOR) adsorption, which was found to be significant. A detailed experimental parametric study enabled the authors to conclude that the adsorption mechanism is mainly influenced by pore filling, electrostatic interaction, and the H-bond.

Funding: The Operational Program "Competitiveness, Entrepreneurship and Innovation", under the call "RESEARCH-CREATE-INNOVATE" (project code: T1EΔK-00782) co-financed by the European Union and Greek national funds, are gratefully acknowledged by I.V.Y.

Acknowledgments: We would like to thank the Nanomaterials editorial team, particularly Assistant Editor Sandra Ma, for their continuous and close support. Thanks to all contributed authors and reviewers for their valuable involvement on the high quality and standards contributing papers produced that can give advances for researchers on the field.

Conflicts of Interest: The authors declare no conflict of interest.

References

1. Lambert, R.M.; Williams, F.J.; Cropley, R.L.; Palermo, A. Heterogeneous alkene epoxidation: past, present and future. *J. Mol. Catal. A Chem.* **2005**, *228*, 27–33. [CrossRef]
2. Huang, Z.; Gu, X.; Cao, Q.; Hu, P.; Hao, J.; Li, J.; Tang, X. Catalytically active single-atom sites fabricated from silver particles. *Angew. Chem. Int. Ed.* **2012**, *51*, 4198–4203. [CrossRef]
3. Yentekakis, I.V.; Goula, G. Biogas Management: Advanced Utilization for Production of Renewable Energy and Added-value Chemicals. *Front. Environ. Sci.* **2017**, *5*, 7. [CrossRef]
4. Yentekakis, I.V.; Papadam, T.; Goula, G. Electricity production from wastewater treatment via a novel biogas-SOFC aided process. *Solid State Ion.* **2008**, *179*, 1521–1525. [CrossRef]
5. Papadam, T.; Goula, G.; Yentekakis, I.V. Long-term operation stability tests of intermediate and high temperature Ni-based anodes' SOFCs directly fueled with simulated biogas mixtures. *Int. J. Hydr. Energy* **2012**, *37*, 16680–16685. [CrossRef]

6. Mao, Y.; Zhang, X.; Zhou, Y.; Chu, W. Microwave-assisted synthesis of porous nano-sized $Na_3V_2(PO_4)_2F_3$@C nanospheres for sodium ion batteries with enhanced stability. *Scr. Mater.* **2020**, *181*, 92–96. [CrossRef]
7. Wei, M.; Li, J.; Chu, W.; Wang, N. Phase control of 2D binary hydroxides nanosheets via controlling-release strategy for enhanced oxygen evolution reaction and supercapacitor performances. *J. Energy Chem.* **2019**, *38*, 26–33. [CrossRef]
8. Yentekakis, I.V.; Goula, G.; Hatzisymeon, M.; Betsi-Argyropoulou, I.; Botzolaki, G.; Kousi, K.; Kondarides, D.I.; Taylor, M.J.; Parlett, C.M.A.; Osatiashtiani, A.; et al. Effect of support oxygen storage capacity on the catalytic performance of Rh nanoparticles for CO_2 reforming of methane. *Appl. Catal. B* **2019**, *243*, 490–501. [CrossRef]
9. Yentekakis, I.V.; Vernoux, P.; Goula, G.; Caravaca, A. Electropositive Promotion by Alkalis or Alkaline Earths of Pt-Group Metals in Emissions Control Catalysis: A Status Report. *Catalysts* **2019**, *9*, 157. [CrossRef]
10. Papavasiliou, A.; Tsetsekou, A.; Matsuka, V.; Konsolakis, M.; Yentekakis, I.V.; Boukos, N. Synergistic structural and surface promotion of monometallic (Pt) TWCs: Effectiveness and thermal aging tolerance. *Appl. Catal. B* **2011**, *106*, 228–241. [CrossRef]
11. Wang, Y.; Pan, C.; Chu, W.; Vipin, A.K.; Sun, L. Environmental Remediation Applications of Carbon Nanotubes and Graphene Oxide: Adsorption and Catalysis. *Nanomaterials* **2019**, *9*, 439. [CrossRef] [PubMed]
12. Yentekakis, I.V.; Vernoux, P. Emissions Control Catalysis. *Catalysts* **2019**, *9*, 912. [CrossRef]
13. Yentekakis, I.V.; Goula, G.; Kampouri, S.; Betsi-Argyropoulou, I.; Panagiotopoulou, P.; Taylor, M.J.; Kyriakou, G.; Lambert, R.M. Ir-catalyzed Nitrous oxide (N_2O) decomposition: Effect of the Ir particle size and metal-support interactions. *Catal. Lett.* **2018**, *148*, 341–347. [CrossRef]
14. Yang, X.-F.; Wang, A.; Qiao, B.; Li, J.; Liu, J.; Zhang, T. Single-atom catalysts: A new frontier in heterogeneous catalysis. *Acc. Chem. Res.* **2013**, *46*, 1740–1748. [CrossRef] [PubMed]
15. Flytzani-Stephanopoulos, M.; Gates, B.C. Atomically dispersed supported metal catalysts. *Ann. Rev. Chem. Biomol. Eng.* **2012**, *3*, 545–574. [CrossRef]
16. Datye, A.; Wang, Y. Atom trapping: A novel approach to generate thermally stable and regenerable single-atom catalysts. *Natl. Sci. Rev.* **2018**, *5*, 630–632. [CrossRef]
17. Vernoux, P.; Lizarraga, L.; Tsampas, M.N.; Sapountzi, F.M.; De Lucas-Consuegra, A.; Valverde, J.-L.; Souentie, S.; Vayenas, C.G.; Tsiplakides, D.; Balomenou, S.; et al. Ionically Conducting Ceramics as Active Catalyst Supports. *Chem. Rev.* **2013**, *113*, 8192–8260. [CrossRef]
18. Goula, G.; Botzolaki, G.; Osatiashtiani, A.; Parlett, C.M.A.; Kyriakou, G.; Lambert, R.M.; Yentekakis, I.V. Oxidative Thermal Sintering and Redispersion of Rh Nanoparticles on Supports with High Oxygen Ion Lability. *Catalysts* **2019**, *9*, 541. [CrossRef]
19. Su, P.; Liu, X.; Chen, Y.; Liu, H.; Zhu, B.; Zhang, S.; Huang, W. Synthesis and Characterization of Rh/B–TNTs as a Recyclable Catalyst for Hydroformylation of Olefin Containing–CN Functional Group. *Nanomaterials* **2018**, *8*, 755. [CrossRef]
20. Riyadh, S.; Khalil, K.; Aljuhani, A. Chitosan-MgO Nanocomposite: One Pot Preparation and Its Utility as an Ecofriendly Biocatalyst in the Synthesis of Thiazoles and [1,3,4] thiadiazoles. *Nanomaterials* **2018**, *8*, 928. [CrossRef]
21. Zhao, L.; Zhao, J.; Wu, T.; Zhao, M.; Yan, W.; Zhang, Y.; Li, H.; Wang, Y.; Xiao, T.; Zhao, Y. Synergistic Effect of Oxygen Vacancies and Ni Species on Tuning Selectivity of Ni/ZrO_2 Catalyst for Hydrogenation of Maleic Anhydride into Succinic Anhydride and γ-Butyrolacetone. *Nanomaterials* **2019**, *9*, 406. [CrossRef] [PubMed]
22. Wu, J.; Wang, S.; Li, H.; Zhang, Y.; Shi, R.; Zhao, Y. The Synergistic Effect of Acidic Properties and Channel Systems of Zeolites on the Synthesis of Polyoxymethylene Dimethyl Ethers from Dimethoxymethane and Trioxymethylene. *Nanomaterials* **2019**, *9*, 1192. [CrossRef] [PubMed]
23. Chatzikonstantinou, A.; Gkantzou, E.; Thomou, E.; Chalmpes, N.; Lyra, K.; Kontogianni, V.; Spyrou, K.; Patila, M.; Gournis, D.; Stamatis, H. Enzymatic Conversion of Oleuropein to Hydroxytyrosol Using Immobilized β-Glucosidase on Porous Carbon Cuboids. *Nanomaterials* **2019**, *9*, 1166. [CrossRef]
24. Liu, H.; Zhang, Z.; He, H.; Wang, X.; Zhang, J.; Zhang, Q.; Tong, Y.; Liu, H.; Ramakrishna, S.; Yan, S.; et al. One-Step Synthesis Heterostructured g-C_3N_4/TiO_2 Composite for Rapid Degradation of Pollutants in Utilizing Visible Light. *Nanomaterials* **2018**, *8*, 842. [CrossRef] [PubMed]
25. Bresolin, B.; Ben Hammouda, S.; Sillanpää, M. An Emerging Visible-Light Organic–Inorganic Hybrid Perovskite for Photocatalytic Applications. *Nanomaterials* **2020**, *10*, 115. [CrossRef] [PubMed]

26. Chen, J.; Huang, S.; Long, Y.; Wu, J.; Li, H.; Li, Z.; Zeng, Y.; Ruan, S. Fabrication of ZnO/Red Phosphorus Heterostructure for Effective Photocatalytic H_2 Evolution from Water Splitting. *Nanomaterials* **2018**, *8*, 835. [CrossRef]
27. Wen, Y.; Wei, Z.; Ma, C.; Xing, X.; Li, Z.; Luo, D. MXene Boosted CoNi-ZIF-67 as Highly Efficient Electrocatalysts for Oxygen Evolution. *Nanomaterials* **2019**, *9*, 775. [CrossRef]
28. Wang, H.; Zhang, X.; Wang, Y.; Quan, G.; Han, X.; Yan, J. Facile Synthesis of Magnetic Nitrogen-Doped Porous Carbon from Bimetallic Metal–Organic Frameworks for Efficient Norfloxacin Removal. *Nanomaterials* **2018**, *8*, 664. [CrossRef]

nanomaterials

MDPI

Article

Synthesis and Characterization of Rh/B–TNTs as a Recyclable Catalyst for Hydroformylation of Olefin Containing –CN Functional Group

Penghe Su [1], Xiaotong Liu [1], Ya Chen [1], Hongchi Liu [1], Baolin Zhu [1,2], Shoumin Zhang [1,2] and Weiping Huang [1,2,3,*]

1 College of Chemistry, Nankai University, Tianjin 300071, China; sph_edu@163.com (P.S.); jenalia@163.com (X.L.); xiaoyayaking@163.com (Y.C.); h335liu@edu.uwaterloo.ca (H.L.); zhubaolin@nankai.edu.cn (B.Z.); zhangsm@nankai.edu.cn (S.Z.)

2 The Key Laboratory of Advanced Energy Materials Chemistry (Ministry of Education), Nankai University, Tianjin 300071, China

3 Collaborative Innovation Center of Chemical Science and Engineering, Tianjin 300071, China

* Correspondence: hwp914@nankai.edu.cn; Tel.: +86-138-2009-6974

Received: 7 September 2018; Accepted: 20 September 2018; Published: 25 September 2018

Abstract: The TiO_2-based nanotubes (TNTs, B–TNTs) of different surface acidities and their supported Rh catalysts were designed and synthesized. The catalysts were characterized by X-ray diffraction (XRD), scanning electron microscopy (SEM), transmission electron microscopy (TEM), X-ray photoelectron spectrometer (XPS), tempera–ture–programmed desorption of ammonia (NH_3–TPD), atomic emission spectrometer (ICP), and Brunauer–Emmett–Tellerv (BET) surface-area analyzers. Images of SEM and TEM showed that the boron-decorated TiO_2 nanotubes (B–TNTs) had a perfect multiwalled tubular structure; their length was up to hundreds of nanometers and inner diameter was about 7 nm. The results of NH_3-TPD analyses showed that B–TNTs had a stronger acid site compared with TNTs. For Rh/TNTs and Rh/B–TNTs, Rh nanoparticles highly dispersed on B–TNTs were about 2.79 nm in average diameter and much smaller than those on TNTs, which were about 4.94 nm. The catalytic performances of catalysts for the hydroformylation of 2-methyl-3-butennitrile (2M3BN) were also evaluated, and results showed that the existence of B in Rh/B–TNTs had a great influence on the catalytic performance of the catalysts. The Rh/B–TNTs displayed higher catalytic activity, selectivity for aldehydes, and stability than the Rh/TNTs.

Keywords: B-doped; Rh; TiO_2 nanotube; hydroformylation; 2-methyl-3-butennitrile; functionalized olefin

1. Introduction

Hydroformylation of olefins is one of the most important homogeneous catalytic reactions in the chemical industry with a worldwide oxoaldehyde production [1]. By hydroformylation, one more carbon aldehyde than the original olefin can be obtained. Aldehydes are an important chemical raw material and act as intermediates in the synthesis of drugs, pesticides, natural products, and so on [2–4]. According to statistics, the production of aldehydes globally is now over 6.0×10^6 t/a [5].

The typical hydroformylation of olefins is mainly catalyzed by homogeneous catalysis. Although the homogeneous catalyst system for hydroformylation has high catalytic activity, good selectivity, and other advantages, the transition metal complex used in the catalyst system may dissolve in the product, resulting in difficulties in the recovery of the catalyst [6,7]. Thus, there are two technical and scientific issues; one is the separation of the homogeneous catalyst from the product, and the other is preventing active components from loss. To resolve these issues, researchers have done much research on the effective separation of homogeneous catalysts and refraining from the loss of active components for the hydroformylation of olefins. One effective method is to "heterogenize" homogeneous catalysts. That is, the homogeneous catalyst, or the metal nanoparticles catalyst, is riveted on the materials of a large surface area [8–14]. Compared with conventional inorganic supports, the TiO_2 nanotubes (TNTs) with multi-hydroxyls on the inner and outer surface can support the nanosized catalytic activity center, e.g., Rh nanoparticles. Furthermore, the multiwall of nanotubes may also act as the partition to separate Rh nanoparticles and prevent them from agglomeration. In addition, the nano-confinement effect of TNTs may have a marked impact on the selectivity of hydroformylation. In our previous studies, we used TiO_2 nanotubes-supported Rh nanoparticles and amorphous Co–B catalysts to catalyze the hydroformylations of vinyl acetate [15] and cyclohexene [16]. According to previous reports, the acidity of the supports has a great influence on the catalytic activity. Olefin can be easily adsorbed by the Lewis acid positions [17,18], which is beneficial for the hydroformylation of olefin. The addition of B in the catalyst should be able to improve the catalytic activity of catalysts by increasing the Lewis acid positions [19,20]. Then, we used boron-modified TiO_2 nanotubes-supported Rh-nanoparticle catalysts to catalyze the hydroformylations of styrene, and the TOF of aldehydes was up to 18,458 h^{-1} [21].

The hydroformylation of functionalized alkenes is an interesting topic [22]. The functional group (FG) in functionalized alkenes (C=C–FG), in which the FG is adjacent to the C=C group, may affect catalytic hydroformylation of C=C–FG by the chelation effect of FG with the active sites of the catalyst, which may lead to a decrease in the catalytic activity of the catalyst and the object control of regioselectivity of the catalytic hydroformylation of C=C–FG becomes very difficult. To reduce the impact of FG, we reported TNTs-supported Rh-Ru particle catalysts and compared their catalytic performances in the hydroformylation of vinyl acetate and cyclohexene [23]. The catalysts showed higher catalytic activities in the hydroformylation of vinyl acetate than that of cyclohexene because in the reaction, the main active site, Rh, can catalyze the main reaction efficiently, and the second active site, Ru, reduces the influence of the FG. Olefins containing the –CN group are important functionalized alkenes. Scientists have tried hard to regulate and control the regioselectivity of the catalytic hydroformylation of olefins containing the –CN functional group [24–26]. It is well-known that the control addition of two HCN to a single butadiene is still the most effective industrial process to synthesize adiponitrile today. When one molecule of HCN is added to a butadiene, the main byproduct is 2M3BN. 2M3BN, as is commonly known, can be used for the preparation of adiponitrile after isomerization, but investigating the hydroformylation of 2M3BN is an industrially important and scientifically challenging research subject. In this contribution, we report the design and synthesis of Rh-catalysts supported by TNTs of different surface acidity. The catalytic performances of catalysts synthesized for the hydroformylation of 2M3BN are investigated.

2. Materials and Methods

Butyl titanate, ethanol, H_3BO_3, NaOH, and nitric acid are analytical grade reagents and were purchased from commercial suppliers (Tianjin Guangfu Fine Chemical Research Institute, Tianjin, China) without any further purification. Deionized water is used in the experiments.

2.1. Preparation of the Catalysts

The preparation method of the TNTs was the same as our previous studies [27].

The preparation method of the B–TNTs is described as follows. Firstly, 15 mL butyl titanate was added into 225 mL absolute ethanol under constant stirring, marked as A. 1.01 g H_3BO_3 was dissolved in 15 mL distilled water respectively, and then the solution was slowly added to 225 mL absolute ethanol, marked as B. Solution A was mixed with solution B, and the pH was adjusted to 3 by using concentrated nitric acid, forming titanium sol containing 5 wt.% boron. After the titanium sols were aged for 1 day, it was dried at 80 °C and turned into titanium gel containing boron. Then, the gel was calcined for 2 h at 400 °C in a muffle furnace, and the boron-doped titanium dioxide powder was obtained. Thereafter, the as-prepared 1.0 g boron-doped titanium dioxide powder was treated with 60 mL 10 M NaOH aqueous solution in an autoclave for 12 h at 150 °C. The resulting material was washed with distilled water for it to become neutral and then dried at 60 °C, whereby a resulting white B–TNTs powder was obtained. The titanium sols containing 10 wt.% boron can be prepared by changing the amount of H_3BO_3 to 2.02 g. The actual B contents of the two B–TNTs powders synthesized and labeled as B–TNTs (b) and B–TNTs (c) were 0.56 and 0.99 wt.%, respectively.

2.2. Synthesis of Rh/TNTs and Rh/B–TNTs

The Rh/TNTs and Rh/B–TNTs were prepared by using the impregnation-photoreducing procedure. A typical synthesis process is as follows: 1 g TNTs was added to 20 mL 2 wt.% RhAc2 aqueous solution, and then the suspension was vigorously agitated for 6 h. After low-energy sonication for 30 min, the suspension was centrifuged and the resulting solid was washed with a little deionized water to remove the ions adsorbed on the outer surface of TNTs, after which the solid was transferred into a 60 mL photo-reactor with 50 mL ethanol–water solution (Vethanol:Vwater = 9:1). The mixture was irradiated using a 300 W high-pressure mercury lamp, which ended with a color change of the mixture from blue to grey. Then, the mixture was centrifuged and the solid obtained was washed with deionized water and dried at 60 °C for 8 h in a vacuum. The catalyst obtained was labeled as Rh/TNTs (a_1). The same procedures were used to prepare Rh/B–TNTs (b_1), Rh/B–TNTs (c_1) to just change TNTs to B–TNTs (b) and B–TNTs (c). The preparation process of Rh/B–TNTs (c_2) and Rh/B–TNTs (c_3) is similar to that of Rh/B–TNTs (c_1), which had different Rh loadings (1 wt.% and 3 wt.% RhAc2).

2.3. Evaluation of Catalytic Performance of Catalysts for Hydroformylation

The catalytic activities of Rh/TNTs and Rh/B–TNTs for the hydroformylation of 2M3BN were evaluated. In a typical experiment, a 0.4 g catalyst and the required amount of substrate and solvent were placed in a 250-mL stainless steel autoclave reactor with a magnetic stirrer. The reactor was purged three times with H_2 and then pressurized to 6.0 MPa with CO and H_2 (CO/H_2 = 1:1). After this, the reactor was heated to the desired temperature while stirring. When the reaction was over, the stirring was stopped. The reactor was then cooled to room temperature and the pressure was released gradually. The product was analyzed by GC7890B-5977A MS (Agilent, Santa Clara, CA, USA) or GC (GC-2014 gas chromatograph equipped with a 30 m × 0.53 mm SE-30 capillary column and a FID, Shimadzu, Japan,). Recycling uses of catalysts were carried out under the same conditions after recovering the catalyst from the reaction solution via centrifugation.

2.4. Characterization

The crystal phase and structure of catalysts were detected by XRD, Rigaku D/Max-2500, (Rigaku, Japan), which was performed with Cu Ka radiation (λ = 1.54 Å) at 2θ from 10° to 80°. TEM images were recorded using a Tecnai G2 F20 instrument (FEI, Hillsboro, OR, USA) at an accelerating voltage of 200 kV. The morphologies were analyzed by SEM, Shimadzu SS-550, Shimadzu, Japan,). The X-ray photoelectron spectrometer (XPS, Kratos Axis Ultra DLD multi-technique X-ray photoelectron spectra, Kratos Analytical Ltd., Manchester, UK) was used to test the chemical states of Rh in catalysts, and all binding energies were calibrated using C1s (E_b = 284.6 eV) for the reference. The contents of Rh and B were determined by ICP-AES (ICP-9000, USA Thermo Jarrell-Ash Corp, Franklin, MA, USA). N_2 adsorption/desorption isotherms were collected on an Autosorb-1-MP 1530VP (Quantachrome, Florida, FL, USA) automatic surface area and porosity analyzer. The sample was degassed at 473 K for 5 h and then analyzed at 77 K. The relative pressure (P/P_0) range used for the calculation of the BET surface area was from 0.05 to 0.30.

NH_3-TPD measurement was performed on an Autosorb-IQ-C-XR automatic surface area and a porosity analyzer equipped with a thermal conductivity detector (TCD, Quantachrome, Florida, FL, USA). The samples were pretreated at 150 °C for 1 h in a flow of Ar, and then subjected to NH_3 adsorption until saturation at room temperature. Desorbed NH_3 was monitored at a heating rate of 10 °C/min from room temperature to 800 °C.

3. Results and Discussion

3.1. BET and ICP Analysis

A Brunauer–Emmett–Teller (BET) analysis was carried out to investigate the specific surface area (SSA) of the catalysts (Table 1). The actual contents of Rh and B in catalysts determined by the ICP were listed in Table 1. The SSA of TNTs (a), B–TNTs (b), and B–TNTs (c) were calculated to be 227.6, 238.6, and 275.8 m^2/g, respectively. It can be observed that the BET surface area increased with the increasing content of B, an effect which can be attributed to the lower crystal size of the anatase phase in the B–TNTs supporter [28,29]. The SSA of all catalysts were significantly lower than that of pure supporters, which may be ascribed to the metal nanoparticles deposited on the outer and inner surface or which occupied the interspace between walls of TNTs [23,30–32].

Table 1. Brunauer–Emmett–Teller (BET) and atomic emission spectrometer (ICP) data of TNTs, Rh/TNTs, B–TNTs and Rh/B–TNTs.

Entry	Catalyst	SSA (m^2/g)	Rh Content (wt.%)	B Content (wt.%)
1	TNTs(a)	227.6	-	-
2	Rh/TNTs(a_1)	198.6	0.13	-
3	B–TNTs(b)	238.6	-	0.56
4	Rh/B–TNTs(b_1)	225.2	0.16	0.56
5	B–TNTs (c)	275.8	0	0.99
6	Rh/B–TNTs(c_1)	268.6	0.16	0.99
7	Rh/B–TNTs(c_2)	270.2	0.09	0.99
8	Rh/B–TNTs(c_3)	267.4	0.19	0.99

3.2. XRD Analysis

Figure 1 shows the XRD patterns of Rh/TNTs, Rh/B–TNTs. It can be seen from Figure 1 that all of the peaks can be perfectly indexed to anatase titania (PDF: 21-1272). The peaks at $2\theta = 25.4°$, 37.8°, 48.1°, 54.6°, 55.0°, 62.7°, 68.8°, and 75.0° can be assigned to diffractions of anatase TiO_2 (101), (004), (200), (105), (211), (204), (116), and (215). There is no peak related to the Rh (PDF: 05-0685), which means that Rh particles are small and highly dispersed on the TNTs and B–TNTs, which is a similar result to that observed by previous reports [15,23].

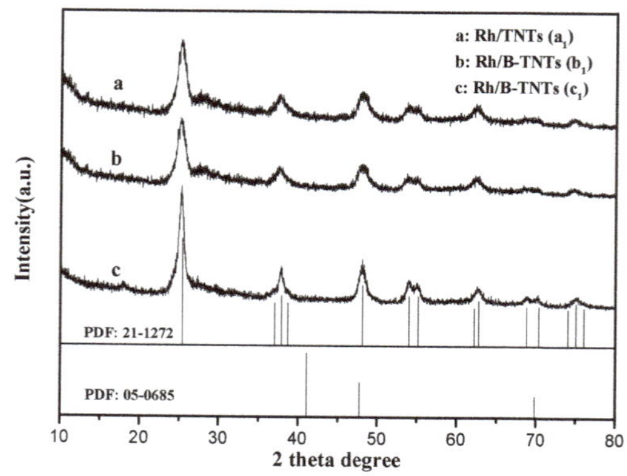

Figure 1. X-ray diffraction (XRD) patterns of Rh/TNTs and Rh/B–TNTs.

3.3. SEM Analysis

Figure 2 shows the SEM images of the B–TNTs. As can be seen from Figure 2, the synthesized B–TNTs showed 1-D morphology with a length of up to hundreds of nanometers. We can see that the single nanotube in the B–TNTs samples is clearly discernible, showing a uniform diameter and a smooth surface.

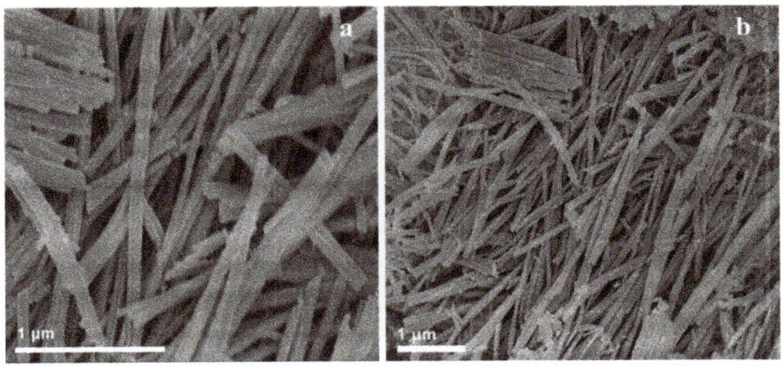

Figure 2. Scanning electron microscopy (SEM) images of a B–TNTs (**a**) and b B–TNTs (**b**).

3.4. TEM Analysis

Figure 3 shows the TEM images of the Rh/TNTs and Rh/B–TNTs. It is clear from Figure 3 that TNTs and B–TNTs show perfect nanotubular morphology with a length of about 200–300 nm and diameter of about 8–10 nm (Figure 3a,b). The nanotubes have a multiwalled structure with a spacing of about 7 nm. There is no obvious rupture or fracture caused by the light reduction process. These are attributable to the dispersed Rh-containing compound, which plays the same stable supporting role as Co and Fe compounds reported in the literature [16,27]. Rh nanoparticles are well-dispersed on the inner and outer surfaces of the nanotubes without agglomeration. Figure 3c,d show the size distributions of particles in different catalysts. The average diameter of the Rh particles in Rh/B–TNTs (c_1) is about 2.79 nm (Figure 3d), which is obviously smaller than that of Rh/TNTs (a_1) (4.94 nm, Figure 3c), and the size distributions of particles are in a relatively narrow range. These results mean that the presence of B in B–TNTs can effectively prevent the growth of Rh nanoparticles and improve the distributions of Rh particles, which is similar to the results found in literature [33].

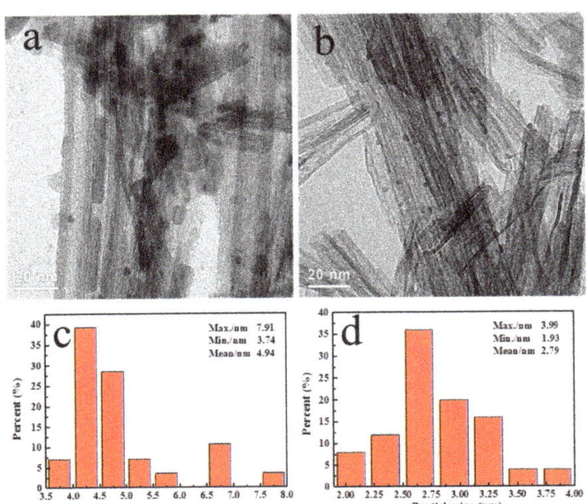

Figure 3. Transmission electron microscopy (TEM) images of (**a**) Rh/TNTs (a_1); (**b**) Rh/B–TNTs (c_1); Size distributions of particles in (**c**) Rh/TNTs (a_1); (**d**) Rh/B–TNTs (c_1).

3.5. XPS Analysis

XPS was performed to evaluate the chemical state of the catalysts. As seen in Figure 4a, there was only one peak at about 192.4 eV for B 1s. As compared to the standard binding energy for B 1s in TiB_2 (187.5 eV, Ti–B bonds) and in B_2O_3 (193.1 eV, B–O bonds), the binding energy of B 1s was between that of B_2O_3 and TiB_2; thus, the boron atom was probably incorporated into TiO_2 and the chemical environment surrounding the boron was likely to be Ti–B–O, which would be a similar result to those in previous reports [34,35].

Figure 4b,c display the O 1s spectra comparison between Rh/TNTs (a_1) and Rh/B–TNTs (c_1). From Figure 4b,c, the O 1s XPS spectra of the Rh/B–TNTs (c_1) sample could be fitted by three peaks, corresponding to Ti–O (530.0 eV), the hydrogen group O–H (531.6 eV), and the B–O (533.0 eV) bond, respectively, while there were only two O 1s XPS peaks assigned to the Ti–O and O–H bonds, respectively of the Rh/TNTs (a_1) sample, which also proved the presence of the B–O bond [35,36].

Figure 4d,e show that the carbon peaks consisted of three components. The peak at 284.6 eV corresponds to the binding energies of C 1s, and the second and third smaller peaks at 286.2 and 288.6 eV observed for both samples are probably due to impurity and air absorbents [37].

As shown in Figure 4f,g, Rh $3d_{5/2}$ and Rh $3d_{3/2}$ show that the group peaks not only centered at 307.2 and 312.0 eV, but also at 309.1 and 313.8 eV. These indicate that Rh exists in two forms of Rh^0 and oxidation state. The higher binding energy state should have relevance to the incomplete reduction of Rh^{2+} ions; however, the electron transfers from Rh adatoms to the titanate substrate and the formation of partially oxidized states cannot be excluded, which is consistent with the literature [38]. It is well-known that Rh^0 is the active site for the hydroformylation of olefins [15]. The percentage of Rh^0 in rhodium is calculated based on the area of the fitted Gauss peaks. When the Rh loading is almost identical, the proportions of Rh^0 in Rh/B–TNTs (c_1) is 89.3%, which is higher than that in Rh/TNTs (a_1). Thus, the incorporation of B can increase the proportion of Rh^0 in the catalyst, promoting the hydroformylation of 2M3BN.

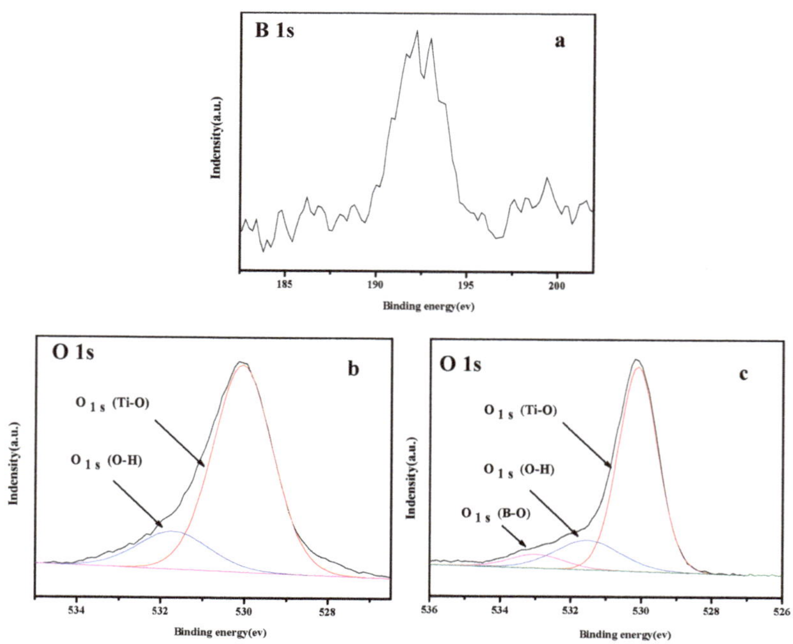

Figure 4. *Cont.*

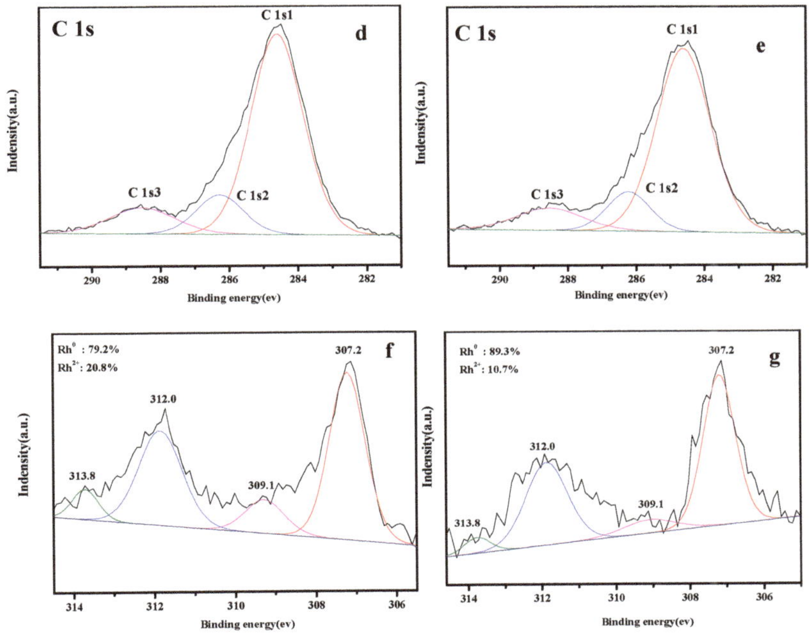

Figure 4. B 1s XPS spectrum of (**a**) Rh/B–TNTs (c_1); O 1s XPS spectra of (**b**) Rh/TNTs (a_1) and (**c**) Rh/B–TNTs (c_1); C 1s XPS spectra of (**d**) Rh/TNTs (a_1) and (**e**) Rh/B–TNTs (c_1); XPS spectra of Rh in catalysts. (**f**) Rh/TNTs (a_1), (**g**) Rh/B–TNTs (c_1).

3.6. NH_3-TPD Analysis

It has been reported that Lewis acid promoted CO insertion which favors hydroformylation over hydrogenation of alkenes [39]. Previous literature [40] reported that the incorporation of boron into Ti-substituted silicalite-2 zeolites (TS-2) could enhance both the number and strength of the acidic sites in TS-2.

NH_3-TPD was used to measure the acidic properties of TNTs (a) and B–TNTs (c), and the results are shown in Figure 5. There are two NH_3 desorption peaks at 135 and 335 °C in the TNTs (Figure 5a), which are caused by the NH_3 adsorbed on the acidic sites of the outer and inner surfaces of TNTs, respectively. It can be concluded that there are a lot of acidic sites on the inner and outer surfaces of TNTs, which is beneficial to the preferential adsorption of olefin to catalysts. Compared with TNTs (a), the B–TNTs (c) show three NH_3 desorption peaks at around 110, 375, and 560 °C (Figure 5b), which is similar to the NH_3-TPD results of Co-B/TNTs [41]. The peak at around 560 °C corresponding to NH_3 adsorbed on stronger acidic sites implies that B may lead to the formation of a new kind of acidic site and enhance the acidity of catalysts.

3.7. Catalytic Activity Evaluation

Scheme 1 shows the possible aldehydes formed in the hydroformylation reaction of 2M3BN.

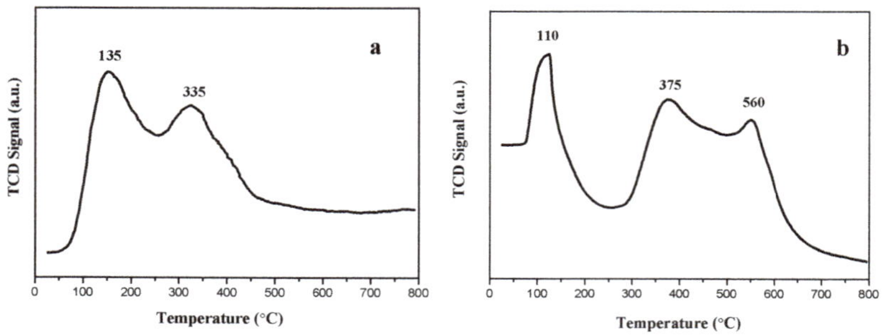

Figure 5. NH_3-TPD plots of a TNTs (**a**) and (**b**) B–TNTs.

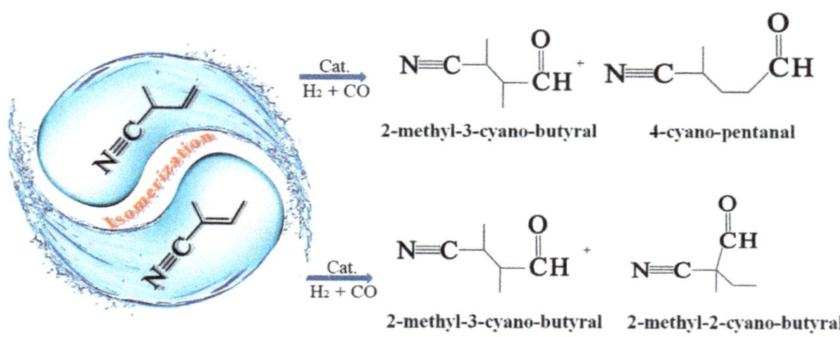

Scheme 1. Possible aldehydes in the hydroformylation reaction of 2M3BN.

In the real hydroformylation reaction of 2M3BN, an isomerization reaction of 2-methyl-2-butennitrile (2M2BN) may take place. Theoretically speaking, the products should include at least three kinds of aldehyde in the hydroformylation of 2M3BN, one of which is linear aldehyde and the others are branched-chain aldehydes (Figures S1–S3).

It is very interesting to study the selectivity for these aldehydes in the hydroformylation of 2M3BN and it can help us to understand how the –CN functional group affects the reaction process of the hydroformylation of olefin. The GC and MS fragments of products formed in the hydroformylation of 2M3BN are shown in the supporting information.

Table 2 shows the influence of B content on the activity of the catalyst. As can be seen from Table 2, the conversion rate of 2M3BN over all catalysts can reach up to 100%.

However, the selectivity for aldehyde in the reaction over Rh/B–TNTs is higher than that over Rh/TNTs. The selectivity for product aldehydes is shown to increase from 72% to 81%. In addition, the isomerization of 2M3BN clearly reduces from 25.8% to 17.9%. These results imply that in the catalytic processes, the presence of B in the catalysts may increase the rate of migratory insertion of the CO group, which suppresses the isomerization of 2M3BN and improves reaction selectivity for aldehydes. When the amount of B in Rh/B–TNTs increases from 0.56 wt.% to 0.99 wt.%, both of the selectivity for aldehyde and the l:b ratio of the product aldehydes increase. Thus, the suitable amount of B doping should be 0.99 wt.% in the present work.

Table 2. Effect of boron content on the hydroformylation over Rh/B–TNTs [a].

Catalyst	Content of B (wt.%)	Conversion (%)	Isomerism [d] (%)	TOF [b] (h^{-1})	Aldehyde (%)	b:l [c]
Rh/TNTs(a_1)	0	100	25.8	2663	72	32:68
Rh/BTNTs(b_1)	0.56	100	20.4	2532	79	33:67
Rh/BTNTs(c_1)	0.99	100	17.9	3576	81	31:69

[a] Reaction conditions: 2M3BN = 5 mL, catalyst = 0.40 g, temp. = 120 °C, syngas pressure = 6.0 MPa, CO:H_2 = 1, solvent (toluene) = 70 mL, and reaction time = 2 h. [b] TOF is calculated by the formula TOF $= \frac{n(\text{aldehydes})}{n(\text{Rh}) \times \text{h}}$, n(aldehydes) is the total moles of aldehydes, n(Rh) is the total moles of Rh in the catalyst. [c] b:l is branched aldehyde: linear aldehyde. [d] Isomerism $= \frac{n(\text{2-methyl-2-cyano-butyral}) + n(\text{2M2BN})}{n(\text{2M3BN})}$, n(2M2BN) is the amount of 2M2BN in the liquid after reaction, n(2M3BN) is the amount of reaction substrate.

Table 3 shows the effects of Rh content on the catalytic performance of the Rh/B–TNTs. It can be seen from Table 3 that the conversion rate of 2M3BN, the turnover frequency (TOF), and the selectivity for aldehydes all increase with an increase in Rh loadings from 0.09 wt.% to 0.16 wt.%. However, when the amount of Rh increases from 0.16 wt.% to 0.19 wt.%, the selectivity for aldehydes does not increase, but the ratio of linear aldehyde to the branched one clearly decreases. Thus, it can be concluded that the better rhodium content is 0.16wt.% for the catalyst Rh/B–TNTs.

Table 3. Effect of Rh loading in Rh/B–TNTs on the hydroformylation of 2M3BN [a].

Catalyst	Rh Content (wt.%)	Conversion (%)	Isomerism [d] (%)	TOF [b] (h-1)	Aldehyde (%)	b:l [c]
Rh/B–TNTs(c_1)	0.16 (6.2×10^{-3} mmol)	100	17.9	3576	81	31:69
Rh/B–TNTs(c_2)	0.09 (3.5×10^{-3} mmol)	86	19.1	2742	68	32:68
Rh/B–TNTs(c_3)	0.19 (7.3×10^{-3} mmol)	100	18.2	3639	79	42:58

[a] Reaction conditions: 2M3BN = 5 mL, catalyst = 0.40 g, temp. = 120 °C, syngas pressure = 6.0 MPa, CO:H_2 = 1, solvent (toluene) = 70 mL, and reaction time = 2 h. [b] TOF is calculated by the formula TOF $= \frac{n(\text{aldehydes})}{n(\text{Rh}) \times \text{h}}$, n(aldehydes) is the total moles of aldehydes, n(Rh) is the total moles of Rh in catalyst. [c] b:l is branched aldehyde: linear aldehyde. [d] Isomerism $= \frac{n(\text{2-methyl-2-cyano-butyral}) + n(\text{2M2BN})}{n(\text{2M3BN})}$, n(2M2BN) is the amount of 2M2BN in the liquid after reaction, n(2M3BN) is the amount of reaction substrate.

Table 4 shows the effect of reaction temperature on the Rh/TNTs (a_1) and Rh/B–TNTs (c_1) catalyzed hydroformylation reaction of 2M3BN. Table 4 shows that the conversion of 2M3BN over Rh/TNTs, the TOF, and the ratio of l:b, all increased along with the increase in reaction temperature from 80 °C to 120 °C. For Rh/B–TNTs (c_1), when the reaction temperature increased from 80 °C to 120 °C, the conversion of 2M3BN, the total amount of aldehydes, and the ratio of l:b, all improved. Thus, it can be concluded that the suitable reaction temperature should be 120 °C. The difference is that under the almost identical Rh loading and reaction temperature, the activity of Rh/B–TNTs is significantly higher than that of Rh/TNTs. This clearly demonstrates that the presence of B in the catalysts is greatly beneficial for the hydroformylation reaction of 2M3BN. However, the isomerization of 2M3BN increases when the temperature increases from 80 °C to 120 °C, indicating that increasing the temperature is more favorable for the isomerization of 2M3BN.

The stability of catalysts is important for the hydroformylation of olefins for practical application. It was reported that the stability of Ni catalysts can be enhanced by boron doping [42]. Density functional theory calculations also suggested that the presence of boron could improve the stability of the catalyst by selectively blocking the deposition, nucleation, and growth of resilient carbon species of Co catalysts under Fischer-Tropsch synthesis conditions [43].

Table 4. Effect of reaction temperatures on the Rh/TNTs- and Rh/B–TNTs-catalyzed hydroformylation reaction of 2M3BN [a].

Catalyst	Temperature (°C)	Conversion (%)	Isomerism [d] (%)	TOF [b] (h^{-1})	Aldehyde (%)	b:l [c]
Rh/TNTs (a_1)	80	66	19.5	1833	46	65:35
Rh/TNTs (a_1)	100	100	20.4	3152	73	39:61
Rh/TNTs (a_1)	120	100	25.9	3263	72	32:68
Rh/B–TNTs (c_1)	80	74	13.1	1892	59	62:38
Rh/B–TNTs (c_1)	100	100	16.8	3537	81	41:59
Rh/B–TNTs (c_1)	120	100	17.9	3576	81	31:69

[a] Reaction conditions: 2M3BN = 5 mL, Catalyst = 0.40 g, syngas pressure = 6.0 MPa, CO:H_2 = 1, solvent (toluene) = 70 mL, and reaction time = 2 h. [b] TOF is calculated by the formula TOF $= \frac{n(\text{aldehydes})}{n(\text{Rh}) \times \text{h}}$, n(aldehydes) is the total moles of aldehydes, n(Rh) is the total moles of Rh in catalyst. [c] b:l is branched aldehydee: linear aldehyd. [d] Isomerism $= \frac{n(\text{2-methyl-2-cyano-butyral}) + n(\text{2M2BN})}{n(\text{2M3BN})}$, n(2M2BN) is the amount of 2M2BN in the liquid after reaction, n(2M3BN) is the amount of reaction substrate.

To investigate the stability of the catalysts, Rh/TNTs (a_1) and Rh/B–TNTs (c_1) were selected for recycle catalytic experiments. The experimental results are listed in Table 5. As can be seen from this Table, Rh/B–TNTs (c_1) remained highly active and the yield of aldehyde maintained at around 67.8% in the fourth recycles. However, the Rh/TNTs (a_1) shows poor stability. We used ICP to test the rhodium content in the solution after the reaction (listed in Table 6). We can clearly see that the addition of B can significantly reduce the loss of rhodium. The Rh content is about 18.8 PPM in the solution after the first reaction used for the Rh/B–TNTs(c_1) catalyst and is much smaller than that in the solution used for the Rh/TNTs(a_1) catalyst, which is about 42.8 PPM. The experimental results show that boron doping can greatly improve the catalytic stability of the catalyst.

Table 5. Studies of catalyst stability [a].

Catalyst	Cycle Times	Conversion (%)	Isomerism [d] (%)	TOF [b] (h^{-1})	Aldehyde (%)	b:l [c]
Rh/B–TNTs(c_1)	1	100	17.9	3576	81	31:69
Rh/B–TNTs(c_1)	2	100	20.3	3312	77	29:71
Rh/B–TNTs(c_1)	3	92	18.8	2556	80	31:69
Rh/B–TNTs(c_1)	4	90	29.1	2202	68	38:62
Rh/TNTs(a_1)	1	100	25.9	3263	72	32:68
Rh/TNTs(a_1)	2	50	28.6	1665	69	35:65
Rh/TNTs(a_1)	3	trace	30.5	-	trace	-

[a] Reaction conditions: 2M3BN = 5 mL, Rh/TNTs (a_1) = 0.40 g, Rh/B–TNTs (c_1) = 0.40 g, temp. = 120 °C, syngas pressure = 6.0 MPa, CO:H_2 = 1, solvent (toluene) = 70 mL. [b] TOF is calculated by the formula TOF $= \frac{n(\text{aldehydes})}{n(\text{Rh}) \times \text{h}}$, n(aldehydes) is the total moles of aldehydes, n(Rh) is the total moles of Rh in catalyst. [c] b:l is branched aldehyde: linear aldehyde. [d] Isomerism $= \frac{n(\text{2-methyl-2-cyano-butyral}) + n(\text{2M2BN})}{n(\text{2M3BN})}$, n(2M2BN) is the amount of 2M2BN in the liquid after reaction, n(2M3BN) is the amount of reaction substrate.

Table 6. Rhodium content in the solution after the reaction.

Catalyst	Cycle Times	Rh (PPM)
Rh/B–TNTs(c_1)	1	18.8
Rh/B–TNTs(c_1)	2	15.7
Rh/B–TNTs(c_1)	3	13.0
Rh/B–TNTs(c_1)	4	11.2
Rh/TNTs(a_1)	1	42.8
Rh/TNTs(a_1)	2	28.7
Rh/TNTs(a_1)	3	8.6

4. Conclusions

This study showed that the high catalytic activity of B-modified Rh/TNTs catalysts for olefin hydroformylation could be improved by increasing the acid site of the catalyst. The Rh/B–TNTs catalysts not only display good catalytic performance, but also good circulation. The catalysts could be easily recovered and were used up to four times. Although we observed a modest decrease of activity with the recycling of the catalyst, the selectivity decreased quite a bit from 81% to 68%, and Rh leaching was also observed. However, the stability of the catalyst Rh/B–TNTs(c_1) was much better than that of Rh/TNTs(a_1). In conclusion, the introduction of B can not only improve the catalytic activity of the catalyst, but also improve its stability.

Supplementary Materials: The following are available online at http://www.mdpi.com/2079-4991/8/10/755/s1, Figure S1: Gas chromatogram of hydroformylation. **a** before the reaction; **b** after the reaction. Figure S2: GC-Mass profiles of samples from hydroformylation of 2M3BN in toluene. Fragment ions of the 2-methyl-2-cyano-butyral (RT = 9.47 min); b fragment ions of the 2-methyl-3-cyano-butyral (RT = 10.30 min); c fragment ions of the 4-cyano-pentanal (RT = 13.70 min). Figure S3: IR spectra of hydroformylation production of 2M3BN.

Author Contributions: P.S. performed the experiments and wrote the original draft; X.L., Y.C. and H.L. done the work of data processing; B.Z., S.Z. and W.H. conceived the concept. All the authors contributed to the writing of the manuscript.

Funding: This work is supported by the National Natural Science Foundation of China (21373120 and 21071086) and 111 Project (B12015).

Acknowledgments: This research have no external acknowledgments.

Conflicts of Interest: The authors declare no conflict of interest.

References

1. Piras, I.; Jennerjahn, R.; Jackstell, R.; Spannenberg, A.; Franke, R.; Beller, M. A General and Efficient Iridium-Catalyzed Hydroformylation of Olefins. *Angew. Chem.* **2011**, *50*, 280–284. [CrossRef] [PubMed]
2. You, C.; Wei, B.; Li, X.X.; Yang, Y.S.; Liu, Y.; Lv, H.; Zhang, X.M. Rhodium-catalyzed desymmetrization by hydroformylation of cyclopentenes: synthesis of chiral carbocyclic nucleosides. *Angew. Chem. Int. Ed.* **2016**, *55*, 6511–6514. [CrossRef] [PubMed]
3. Dydio, P.; Reek, J.N.H. Supramolecular control of selectivity in hydroformylation of vinyl arenes: easy access to valuable beta-aldehyde intermediates. *Angew. Chem. Int. Ed.* **2013**, *52*, 3878–3882. [CrossRef] [PubMed]
4. Pospech, J.; Fleischer, I.; Franke, R.; Buchholz, S.; Beller, M. Alternative metals for homogeneous catalyzed hydroformylation reactions. *Angew. Chem. Int. Ed.* **2013**, *52*, 2852–2872. [CrossRef] [PubMed]
5. Neves, Â.C.B.; Calvete, M.J.F.; Pinho e Melo, T.M.V.D.; Pereira, M.M. Immobilized catalysts for hydroformylation reactions: A versatile tool for aldehyde synthesis. *Eur. J. Org. Chem.* **2012**, *32*, 6309. [CrossRef]
6. Paganelli, S.; Piccolo, O.; Baldi, F.; Gallo, M.; Tassini, R.; Rancan, M.; Armelao, L. A new biogenerated Rh-based catalyst for aqueous biphasic hydroformylation. *Catal. Commun.* **2015**, *71*, 32–36. [CrossRef]
7. Sun, Q.; Dai, Z.F.; Liu, X.L.; Sheng, N.; Deng, F.; Meng, X.J.; Xiao, F.S. Highly Efficient Heterogeneous Hydroformylation over Rh-Metalated Porous Organic Polymers: Synergistic Effect of High Ligand Concentration and Flexible Framework. *J. Am. Chem. Soc.* **2015**, *137*, 5204–5209. [CrossRef] [PubMed]
8. Wang, Y.Q.; Yan, L.; Li, C.Y.; Jiang, M.; Wang, W.L.; Ding, Y.J. Highly efficient porous organic copolymer supported Rh catalysts for heterogeneous hydroformylation of butenes. *Appl. Catal. A Gen.* **2018**, *551*, 98–105. [CrossRef]
9. Garcia, M.A.S.; Oliveira, K.C.B.; Costa, J.C.S.; Corio, P.; Gusevskaya, E.V.; Santos, E.N.D.; Rossi, L.M. Rhodium Nanoparticles as Precursors for the Preparation of an Efficient and Recyclable Hydroformylation Catalyst. *Chem. Cat. Chem.* **2015**, *7*, 1566–1572. [CrossRef]
10. Khokhar, M.D.; Shukla, R.S.; Jasra, R.V. Hydroformylation of dihydrofurans catalyzed by rhodium complex encapsulated hexagonal mesoporous silica. *J. Mol. Catal. A Chem.* **2015**, *400*, 1–6. [CrossRef]
11. Ma, Y.B.; Fu, J.; Gao, Z.X.; Zhang, L.B.; Li, C.Y.; Wang, T.F. Dicyclopentadiene Hydroformylation to Value-Added Fine Chemicals over Magnetically Separable Fe_3O_4-Supported Co-Rh Bimetallic Catalysts: Effects of Cobalt Loading. *Catalysts* **2017**, *7*, 103. [CrossRef]
12. Chikkali, S.H.; Bellini, R.; De Bruin, B.D.; Vlugt, J.I.V.D.; Reek, J.N.H. Highly selective asymmetric Rh-catalyzed hydroformylation of heterocyclic olefins. *J. Am. Chem. Soc.* **2012**, *134*, 6607–6616. [CrossRef] [PubMed]

13. Sharma, D.; Ganesh, V.; Sakthivel, A. Rhodium incorporated monometallic cobalt hydrotalcite-type materials: Preparation and its applications for the hydroformylation of alkenes. *Appl. Catal. A* **2018**, *555*, 155–160. [CrossRef]
14. Kontkanen, M.-L.; Tuikka, M.; Kinnunen, N.M.; Suvanto, S.; Haukka, M. Hydroformylation of 1-Hexene over Rh/Nano-Oxide Catalysts. *Catalysts* **2013**, *3*, 324–337. [CrossRef]
15. Shi, Y.K.; Hu, X.J.; Zhu, B.L.; Wang, S.R.; Zhang, S.M.; Huang, W.P. Synthesis and characterization of TiO_2 nanotube supported Rh-nanoparticle catalysts for regioselective hydroformylation of vinyl acetate. *RSC Adv.* **2014**, *4*, 62215–62222. [CrossRef]
16. Hu, X.J.; Shi, Y.K.; Zhang, YJ.; Zhu, B.L.; Zhang, S.M.; Huang, W.P. Nanotubular TiO_2-supported amorphous Co–B catalysts and their catalytic performances for hydroformylation of cyclohexene. *Catal. Commun.* **2015**, *59*, 45–49. [CrossRef]
17. Zhou, G.B.; Pei, Y.; Jiang, Z.; Fan, K.N.; Qiao, M.H.; Sun, B.; Zong, B.N. Doping effects of B in ZrO_2 on structural and catalytic properties of Ru/B-ZrO_2 catalysts for benzene partial hydrogenation. *J. Catal.* **2014**, *311*, 393–403. [CrossRef]
18. Urbano, F.J.; Aramendía, M.A.; Marinas, A.; Marinas, J.M. An insight into the Meerwein–Ponndorf–Verley reduction of α,β-unsaturated carbonyl compounds: Tuning the acid–base properties of modified zirconia catalysts. *J. Catal.* **2009**, *268*, 79–88. [CrossRef]
19. Chatterjee, M.; Ishizaka, T.; Suzuki, T.; Suzuki, A.; Kawanami, H. In situ synthesized Pd nanoparticles supported on B-MCM-41: an efficient catalyst for hydrogenation of nitroaromatics in supercritical carbon dioxide. *Green Chem.* **2012**, *14*, 3415. [CrossRef]
20. Yin, A.Y.; Qu, J.W.; Guo, X.Y.; Dai, W.L.; Fan, K.N. The influence of B-doping on the catalytic performance of Cu/HMS catalyst for the hydrogenation of dimethyloxalate. *Appl. Catal. A* **2011**, *400*, 39–47. [CrossRef]
21. Shi, Y.K.; Hu, X.J.; Chen, L.; Lu, Y.; Zhu, B.L.; Zhang, S.M.; Huang, W.P. Boron modified TiO_2 nanotubes supported Rh-nanoparticle catalysts for highly efficient hydroformylation of styrene. *New J. Chem.* **2017**, *41*, 6120. [CrossRef]
22. You, C.; Li, X.X.; Yang, Y.H.; Yang, Y.S.; Tan, X.F.; Li, S.L.; Wei, B.; Lv, H.; Chung, L.W.; Zhang, X.M. Silicon-oriented regio- and enantioselective rhodium-catalyzed hydroformylation. *Nat. Commun.* **2018**, *9*, 2045. [CrossRef] [PubMed]
23. Chuai, H.Y.; Liu, X.T.; Chen, Y.; Zhu, B.L.; Zhang, S.M.; Huang, W.P. Hydroformylation of vinyl acetate and cyclohexene over TiO_2 nanotube supported Rh and Ru nanoparticle catalysts. *RSC Adv.* **2018**, *8*, 12053–12059. [CrossRef]
24. Clark, T.P.; Landis, C.R.; Freed, S.L.; Klosin, J.; Abboud, K.A. Highly Active, Regioselective, and Enantioselective Hydroformylation with Rh Catalysts Ligated by Bis-3,4-diazaphospholanes. *J. Am. Chem. Soc.* **2005**, *127*, 5040–5042. [CrossRef] [PubMed]
25. Nelsen, E.R.; Brezny, A.C.; Landis, C.R. Interception and Characterization of Catalyst Species in Rhodium Bis(diazaphospholane)-Catalyzed Hydroformylation of Octene, Vinyl Acetate, Allyl Cyanide, and 1-Phenyl-1,3-butadiene. *J. Am. Chem. Soc.* **2015**, *137*, 14208–14219. [CrossRef] [PubMed]
26. Le Goanvic, L.; Ternel, J.; Couturier, J.-L.; Dubois, J.-L.; Carpentier, J.-F. Rhodium-Biphephos-Catalyzed Tandem Isomerization–Hydroformylation of Oleonitrile. *Catalysts* **2018**, *8*, 21. [CrossRef]
27. An, H.Q.; Li, J.X.; Zhou, J.; Li, K.R.; Zhu, B.L.; Huang, W.P. Iron-coated TiO_2 nanotubes and their photocatalytic performance. *J. Mater. Chem.* **2010**, *20*, 603–610. [CrossRef]
28. In, S.; Orlov, A.; Berg, R.; García, F.; Pedrosa-Jimenez, S.; Tikhov, M.S.; Wright, D.S.; Lambert, R.M. Effective Visible Light-Activated B-Doped and B,N-Codoped TiO_2 Photocatalysts. *J. Am. Chem. Soc.* **2007**, *129*, 13790–13791. [CrossRef] [PubMed]
29. Quiñones, D.H.; Rey, A.; Álvarez, P.M.; Beltrán, F.J.; Li Puma, G. Boron doped TiO_2 catalysts for photocatalytic ozonation of aqueous mixtures of common pesticides: Diuron, o-phenylphenol, MCPA and terbuthylazine. *Appl. Catal. B Environ.* **2015**, *178*, 74–81. [CrossRef]
30. Berto, T.F.; Sanwald, K.E.; Eisenreich, W.; Gutiérrez, O.Y.; Lercher, J.A. Photoreforming of ethylene glycol over Rh/TiO_2 and Rh/GaN:ZnO. *J. Catal.* **2016**, *338*, 68–81. [CrossRef]
31. László, B.; Baán, K.; Oszkó, A.; Erdőhelyi, A.; Kiss, J.; Kónya, Z. Hydrogen evolution in the photocatalytic reaction between methane and water in the presence of CO_2 on titanate and titania supported Rh and Au catalysts. *Top. Catal.* **2018**, *61*, 875–888. [CrossRef]

32. László, B.; Baán, K.; Varga, E.; Oszkó, A.; Erdőhelyi, A.; Kónya, Z.; Kiss, J. Photo-induced reactions in the CO_2-methane system on titanate nanotubes modified with Au and Rh nanoparticles. *Appl. Catal. B Environ.* **2016**, *199*, 473–484. [CrossRef]
33. Zhao, S.; Yue, H.R.; Zhao, Y.J.; Wang, B.; Geng, Y.C.; Lv, J.; Wang, S.P.; Gong, J.L.; Ma, X.B. Chemoselective synthesis of ethanol via hydrogenation of dimethyl oxalate on Cu/SiO_2: Enhanced stability with boron dopant. *J. Catal.* **2013**, *297*, 142–150. [CrossRef]
34. Zhao, W.; Ma, W.H.; Chen, C.C.; Zhao, J.C.; Shuai, Z.G. Efficient Degradation of Toxic Organic Pollutants with Ni_2O_3/TiO_2-xBxunder Visible Irradiation. *J. Am. Chem. Soc.* **2004**, *126*, 4782–4783. [CrossRef] [PubMed]
35. Lu, N.; Quan, X.; Li, J.Y.; Chen, S.; Yu, H.T.; Chen, G.H. Fabrication of Boron-Doped TiO_2 Nanotube Array Electrode and Investigation of Its Photoelectrochemical Capability. *J. Phys. Chem. C* **2007**, *111*, 11836–11842. [CrossRef]
36. Chen, D.M.; Yang, D.; Wang, Q.; Jiang, Z.Y. Effects of Boron Doping on Photocatalytic Activity and Microstructure of Titanium Dioxide Nanoparticles. *Ind. Eng. Chem. Res.* **2006**, *45*, 4110–4116. [CrossRef]
37. Atashbar, M.Z.; Sun, H.T.; Gong, B.; Wlodarski, W.; Lamb, R. XPS study of Nb-doped oxygen sensing TiO_2 thin films prepared by sol-gel method. *Thin Solid Films* **1998**, *326*, 238–244. [CrossRef]
38. Pótári, G.; Madarász, D.; Nagy, L.; László, B.; Sápi, A.; Oszkó, A.; Kukovecz, A.; Erdőhelyi, A.; Kónya, Z.; Kiss, J. Rh-induced support transformation phenomena in titanate nanowire and nanotube catalysts. *Langmuir* **2013**, *29*, 3061–3072. [CrossRef] [PubMed]
39. Shriver, D.F. Some relationships between metal cluster chemistry and heterogeneous catalysis. *J. Clust. Sci.* **1992**, *3*, 459–467. [CrossRef]
40. Fu, Z.; Yin, D.; Li, Q.; Zhang, L.; Zhang, Y. Synthesis, characterization and catalytic properties of titanium and boron co-substituted silicalite zeolites. *Micropor. Mesopor. Mater.* **1999**, *29*, 351–359. [CrossRef]
41. Shi, Y.K.; Hu, X.J.; Zhu, B.L.; Wang, S.R.; Zhang, S.M.; Huang, W.P. Hydroformylation of 1-octene over nanotubular TiO_2-supported amorphous Co-B catalysts. *Chem. Res. Chin. Univ.* **2015**, *31*, 851–857. [CrossRef]
42. Xu, J.; Chen, L.W.; Tan, K.F.; Borgna, A.; Saeys, M. Effect of boron on the stability of Ni catalysts during steam methane reforming. *J. Catal.* **2009**, *261*, 158–165. [CrossRef]
43. Tan, K.F.; Chang, J.; Borgna, A.; Saeys, M. Effect of boron promotion on the stability of cobalt Fischer–Tropsch catalysts. *J. Catal.* **2011**, *280*, 50–59. [CrossRef]

nanomaterials

Article

Chitosan-MgO Nanocomposite: One Pot Preparation and Its Utility as an Ecofriendly Biocatalyst in the Synthesis of Thiazoles and [1,3,4]thiadiazoles

Sayed M. Riyadh [1,2], Khaled D. Khalil [2,3,*] and Ateyatallah Aljuhani [1]

1 Department of Chemistry, Faculty of Science, Taibah University, Al-Madinah Al-Mounawrah 30002, Saudi Arabia; riyadh1993@hotmail.com (S.M.R.); ateyatallah@hotmail.com (A.A.)
2 Department of Chemistry, Faculty of Science, Cairo University, Giza 12613, Egypt
3 Department of Chemistry, Faculty of Science, Taibah University, Yanbu 46423, Saudi Arabia
* Correspondence: khd.khalil@yahoo.com

Received: 17 October 2018; Accepted: 4 November 2018; Published: 8 November 2018

Abstract: A chitosan-MgO hybrid nanocomposite was prepared using a simple chemical precipitation method and characterized using Fourier transform spectroscopy (FTIR), elemental analysis (EDX), and scanning electron microscopy (SEM). The nanocomposite was served as a powerful ecofriendly basic catalyst under microwave irradiation in the synthesis of two novel series of 5-arylazo-2-hydrazonothiazoles **4a–j** and 2-hydrazono[1,3,4]thiadiazoles **8a–d**, incorporating a sulfonamide group. The structures of the synthesized products were elucidated by spectral data and elemental analyses. Also, their yield percentages were calculated using triethylamine (as a traditional catalyst) and chitosan-MgO nanocomposite (as a green recyclable catalyst) in a comparative study.

Keywords: chitosan-MgO nanocomposite; heterogeneous catalysis; ethylidenethiosemicarbazides; thiazoles; thiadiazoles

1. Introduction

In the past decade, green chemistry has been developed from a variety of research ideas such as atom economy and heterogeneous catalysis in the last few decades leading up to the 1990s, and in several successful attempts and efforts to overcome the problems of chemical pollution and resource depletion. On the other hand, nanosized materials are of great importance due to their large exposed surface area, high absorbability, and high catalytic efficiency. Nanosized magnesium oxides were being multitalented basic catalysts for many organic reactions. Recently, MgO nanoparticles can be used for the synthesis of pyranopyrazoles [1], aminochromenes [2], and dihydropyridines [3] via multicomponent reactions. The main drawbacks to the potential use of MgO nanoparticles as a basic catalyst are their difficult separation and reusability since the utilized catalyst could not recover quantitatively and the pure products should be obtained after extensive purification processes. Chitosan (CS) is a natural polysaccharide that is commercially produced via the alkaline hydrolysis of chitin. Although this biopolymer has many advantages, such as its renewability, biocompatibility, and biodegradability, its utility is limited in its unmodified form. For a long time, chitosan was used as an ecofriendly basic catalyst in some organic reactions [4–7]. However, the major problem of its use is the limited basic properties (weak catalytic activity) and its high swelling properties and gel formation renders its separation and recovery very difficult. To overcome these drawbacks, a chitosan-MgO nanocomposite [8] could be used as a novel basic heterogeneous catalyst in the form of a solid film. A chitosan-MgO nanocomposite is a three-dimensional, cross-linked, polymeric matrix of chitosan (with active NH and OH functional groups) incorporating magnesium oxide nanoparticles (Figure 1) [9].

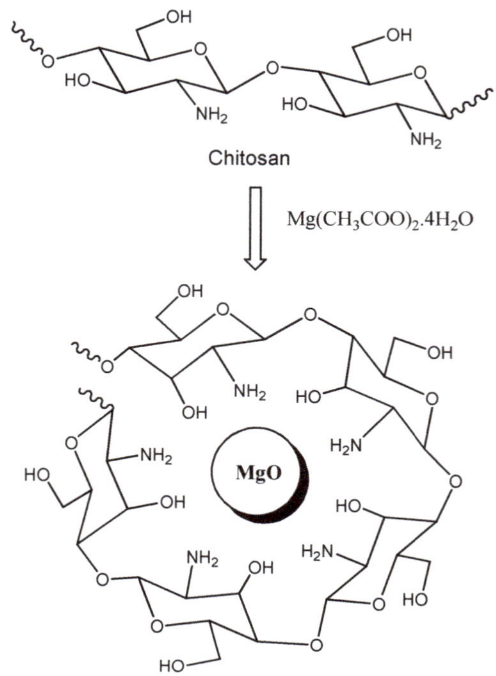

Figure 1. Structure of a chitosan-MgO nanocomposite.

The biocatalyst is stable enough such that it can be recovered and used more than four times without a loss in its catalytic activity. As a result of these studies, the chitosan-MgO biocatalyst was found to be ecofriendly, biocompatible, an efficient green catalyst, and a high economic impact material that will find a lot of potential catalytic applications, especially in the base catalyzed organic transformations. This heterogeneous hybrid nanocomposite is used in the form of films to increase its catalytic activity by increasing its surface/volume ratio, and consequently it can be easily recycled and reused several times with the same catalytic efficiency [10].

Moreover, thiazoles display a diversity of therapeutic activities such as being a non-steroidal anti-inflammatory drug (meloxicam), antiretroviral drug (ritonavir), antiprotozoal agent (nitazoxanide) [11], and antioxidant agent [12]. Also, [1,3,4]thiadiazoles with a sulfonamide group were incorporated in many marketing drugs including acetazolamide and methazolamide [13]. In this study, we are reporting an efficient green protocol and simple synthetic routes of a new series of hydrazonothiazoles and hydrazono [1,3,4]thiadiazoles in the presence of chitosan-MgO nanocomposite as a basic ecofriendly biocatalyst under microwave irradiations. Moreover, the efficiency of this biocatalyst and its reusability was examined.

2. Materials and Methods

Melting points were measured on an electrothermal Gallenkamp capillary apparatus (Leicester, UK) and are uncorrected. Elemental analyses were carried out at the Microanalytical Center of Cairo University, Giza, Egypt. The IR spectra were recorded on a Pye-Unicam SP300 Instrument (Cambridge, UK) in potassium bromide discs. The ^{1}H NMR (nuclear magnetic resonance) and ^{13}C NMR of the newly synthesized compounds in DMSO-d_6 were measured on a Varian Mercury VXR-300 spectrometer (Varian, Karlsruhe, Germany) at 300 MHz for ^{1}H NMR and 75 MHz for ^{13}C NMR) and the chemical shifts were related to that of the solvent. The mass spectra were recorded on a GCMSQ1000-EX Shimadzu (Tokyo, Japan) and GCMS 5988-A HP spectrometers (Shimadzu, Tokyo, Japan) where the ionizing voltage was 70 eV. Microwave experiments were carried out using CEM Discover Labmate microwave apparatus (Discover, SP, NC, USA, 300 W). The starting materials 2-{1-[4-((4-methylphenyl)sulfonamide) phenyl]ethylidine}thiosemicarbazide **1** [14], α-keto hydrazonoyl halides **2a–j** [15–18], and arenecarbohydrazonoyl halides **5a–d** [19,20] were prepared as reported in literature.

2.1. Preparation of Heterogeneous Catalyst (Chitosan-MgO Hybrid Nanocomposite)

The hybrid CS-MgO nanocomposite was prepared using a one pot solution casting method. In an experiment, 2 g of chitosan powder (medium molecular weight; 85% DDA) was dissolved in 100 mL of 2% (v/v) aqueous acetic acid solution under stirring for 12 h at room temperature. To this solution, 2 g of magnesium acetate tetrahydrate, $Mg(CH_3COO)_2 \cdot 4H_2O$ (M0631 Sigma-Aldrich, St. Louis, MO, USA) were added and the mixture was again re-stirred for 12 h till a clear solution was obtained. The resulting viscous solution was cast in a Teflon Petri dish and dried overnight at room temperature. The Petri dish was then immersed into a 0.2 M sodium hydroxide solution for neutralization and phase separation. The solid films were dried in an oven adjusted at 80 °C for 6 h. Finally, the produced solid films were purified by washing with methanol several times and again were dried in an oven at 60 °C for 2 h. Furthermore, a pure chitosan film, without MgO, was prepared in a similar way for comparative study.

2.2. Reactions of 2-{1-[4-((4-Methylphenyl)sulfonamide)phenyl]ethylidine}thiosemicarbazide (1) with α-Keto Hydrazonoyl Chlorides **2a–j** *or N-Aryl Arenecarbohydrazonoyl Halides* **5a–d**

2.2.1. Method A

A mixture of 2-{1-[4-((4-methylphenyl)sulfonamide)phenyl]ethylidine}thiosemicarbazide (1) (0.362 g, 1 mmol) in dry dioxane (15 mL), containing 0.1 g of triethylamine, and α-keto hydrazonoyl chlorides **2a–j** or *N*-aryl arenecarbohydrazonoyl halides **5a–d** (1 mmol of each) was irradiated using microwave irradiation (MW) at 300 W in a closed Teflon vessel until all the starting material was consumed (30–40 min as monitored by thin layer chromatography, TLC). The solvent was evaporated and the residue was poured into an ice/HCl mixture. The precipitate was filtered, washed with water, and crystallized from methanol to give products **4a–j** or **8a–d**.

2.2.2. Method B

The procedure was similar to Method A, using a chitosan-MgO nanocomposite (0.1 g) instead of trimethylamine. After completion of the reaction, the hot solution was filtered to remove chitosan-MgO nanocomposite and the filtrate was poured into an ice/HCl mixture, and the precipitate was filtered, washed with water, and crystallized from methanol to give products **4a–j** or **8a–d** (see Supplementary Materials for analyses of the prepared compounds).

3. Results and Discussion

3.1. Preparation and Characterization of Chitosan-MgO Nanocomposite Films

The chitosan-MgO nanocatalyst was prepared via incorporation of the MgO nanoparticles in the chitosan matrix through a modified one pot solution casting method. The chitosan solution in acetic acid was treated with magnesium acetate tetrahydrate, then was subjected to evaporation of the solvent at room temperature. The resulting solid film was soaked in sodium hydroxide solution to achieve the phase separation.

3.1.1. FTIR Spectra

FTIR (Fourier-Transform infrared) spectra of chitosan and chitosan-MgO nanocomposite showed that the main functional groups of chitosan clearly appeared at υ = 3370 cm^{-1} (broad band of hydrogen bonded OH– groups), 2875 cm^{-1} (C–H bond; CH_3 groups), 1660 cm^{-1} (amide carbonyl groups), 1379 cm^{-1} (bending vibration of CH_2 groups), and 1060 cm^{-1} (asym. vibration of C=O) (Figure 2). These bands are considered as evidence for the maintenance of the chitosan structure features even after the incorporation of the MgO nanoparticles inside the polymer matrix. Also, the absence of acetic acid bands in the spectrum indicated that the films were washed enough and neutralized completely upon sodium hydroxide treatment (see Section 2.1). Moreover, only small acceptable shift in the bands of chitosan-MgO nanocomposite was attributed to the influence of the incorporation of MgO nanoparticles. This shift in bands was familiar as result of chitosan with metal oxides [21]. The latter shifts are shown, especially at bands of NH and OH groups, which is evidence for the H-bonding interaction of these group with MgO molecules.

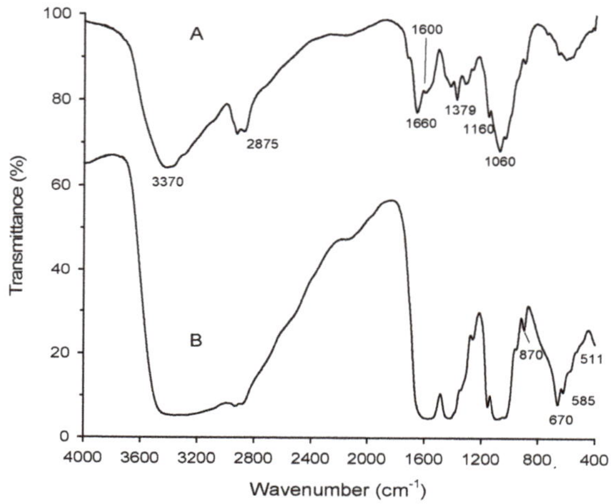

Figure 2. FTIR of chitosan (**A**) and the chitosan-MgO nanocomposite (**B**).

3.1.2. FESEM Analysis

FESEM (Field Emission Scanning Electron Microscope) was used to analyze the morphology and size distribution of the MgO nanoparticles that were incorporated in the polymer matrix. FESEM micrographs of the pure chitosan (A) and that of the hybrid films with magnesia particles 10 wt% (B) are shown in (Figure 3). The obtained surface of the pure chitosan matrix was found to be homogenous and looks smoothly to a great extent. The MgO nanoparticles developed as white spots that distributed homogeneously over the surface of the polymer matrix. On the other hand, the muddled surface was due to adsorption of the polymer layers on the particles surface. The average size of the MgO particles was found to be approximately 6–11 nm for 10 wt% as the particles size slightly decreased with increasing magnesia content. Moreover, EDX measurements for the solid chitosan-MgO nanocomposite confirmed the presence of magnesia within the hybrid film, as shown in (Figure 4).

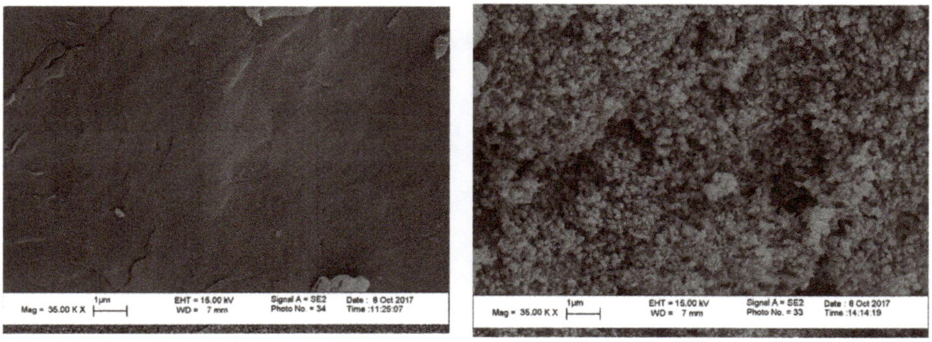

(**A**) (**B**)

Figure 3. FESEM images of chitosan (**A**), and chitosan-MgO 10 wt% (**B**).

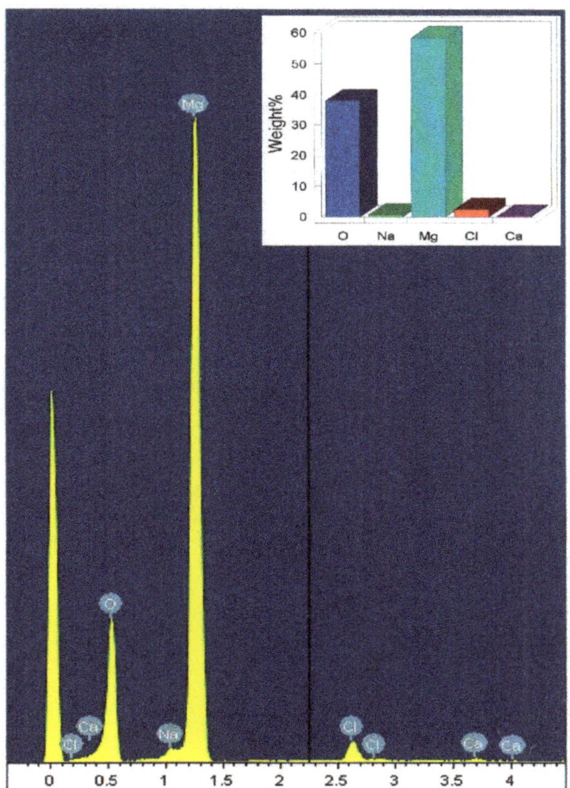

Figure 4. EDX spectra of the 10 wt% chitosan-MgO nanocomposite hybrid film. Inset is the quantitative result.

3.2. *Optimal Catalyst Loading*

In order to estimate the proper catalyst loading, a model irradiation reaction of ethylidinethiosemicarbazide **1** with α-keto hydrazonoyl chlorides **2a** in dioxane, and in the presence of 1, 5, 10, 15, and 20 wt% of chitosan-MgO nanocomposites as basic catalysts, under the same conditions was conducted affording the respective hydrazonothiazoles **4a** (Figure 5). The results showed that the optimal catalyst loading was 10 wt% Moreover, the catalyst was reused four time and the results also showed that the biocatalyst could be reused as such without significant loss in its catalytic activity (Table 1).

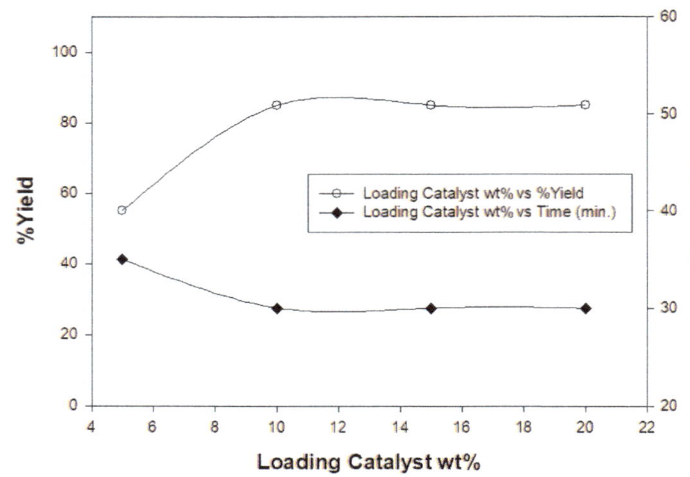

Figure 5. Optimization of the chitosan graft copolymer as a basic catalyst.

Table 1. Recyclability of the chitosan graft copolymer as basic catalyst.

State of Catalyst	Fresh Catalyst	Recycled (1)	Recycled (2)	Recycled (3)	Recycled (4)
Product 4a (%Yield)	85	84	83	83	82

3.3. Synthesis of Thiazoles and [1,3,4]thiadiazoles Using Cs-MgO Nanocomposite

Irradiation of thiosemicarbazone **1** with 2-oxo-*N*-arylpropanehydrazonoyl chlorides **2a–j** in dioxane, in the presence of triethylamine or the chitosan-MgO nanocomposite as a basic catalyst, furnished 4-methyl-5-arylazo-2-hydrazonothiazoles **4a–j** (Scheme 1). At the outset, identification of the best basic catalyst (triethylamine or chitosan-MgO nanocomposite) was examined under microwave irradiations (Table 2).

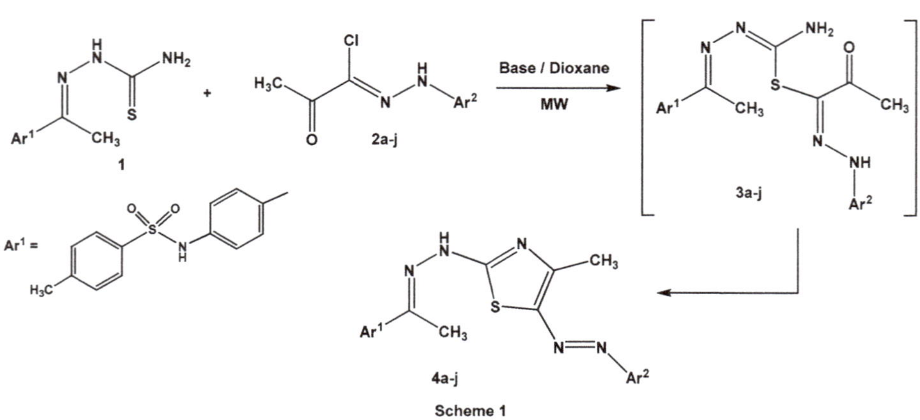

Scheme 1. Synthesis of hydrazonothiazoles **4a–j**.

Table 2. Effect of the nature of basic catalyst on the yield of the products **4a–j**.

Compound Number	Ar^2	Time (min)	Yield (%)	
			Et_3N	Chitosan-MgO Nanocomposite
4a	C_6H_5	30	76	85
4b	2-$CH_3C_6H_4$	30	73	85
4c	2-ClC_6H_4	35	77	86
4d	3-$CH_3C_6H_4$	30	82	96
4e	4-$CH_3C_6H_4$	30	78	87
4f	4-$CH_3OC_6H_4$	40	73	83
4g	4-$NO_2C_6H_4$	40	69	81
4h	4-BrC_6H_4	40	70	81
4i	4-FC_6H_4	40	71	80
4j	4-$CH_3COC_6H_4$	40	72	84

As shown in Table 2, the reaction proceeded smoothly with different substituents on the aromatic benzene ring of hydrazonoyl chlorides **2a–j**. Also, the chitosan-MgO nanocomposite was a more efficient basic catalyst than triethylamine under microwave irradiation.

Elucidation of 2-hydrazonothiazoles **4a–j** structures was based on spectral data and elemental analyses. In the IR spectra, two absorption bands in the range of υ = 3212–3268 cm^{-1} and 1578–1600 cm^{-1} were revealed owing to the presence of (N–H) and (C=N) groups, respectively. Also, the sulfonamide group (SO_2NH) showed asymmetric and symmetric stretching signals at υ = 1336–1379 cm^{-1} and 1136–1161 cm^{-1}, respectively [22]. In ^1H NMR spectra two methyl groups bordering to hydrazone moiety (CH_3–C=N–NH) [23] and a thiazole ring [24] were observed as singlet signals at δ = 2.35–2.43 and 2.43–2.59 ppm, respectively, while the NH proton of the sulfonamide group [22] was resonated at δ = 10.43–10.92 ppm.

Establishing the experimental feasibility of the reaction of **1** with α-keto hydrazonoyl chlorides **2a–j** directed our attention to use *N*-aryl arenecarbohydrazonoyl halides **5a–d**, bereft of the carbonyl group. Thus, treatment of ethylidinethiosemicarbazide **1** with *N*-aryl arenecarbohydrazonoyl halides **5a–d** under the same employed conditions proceeded smoothly to give 2-hydrazono[1,3,4] thiadiazoles **8a–d** as the isolated products (Scheme 2). Also, the effect of the nature of the basic catalyst, such as triethylamine or the chitosan-MgO nanocomposite, on the percent yields of the isolated products **8a–d** was investigated (Table 3).

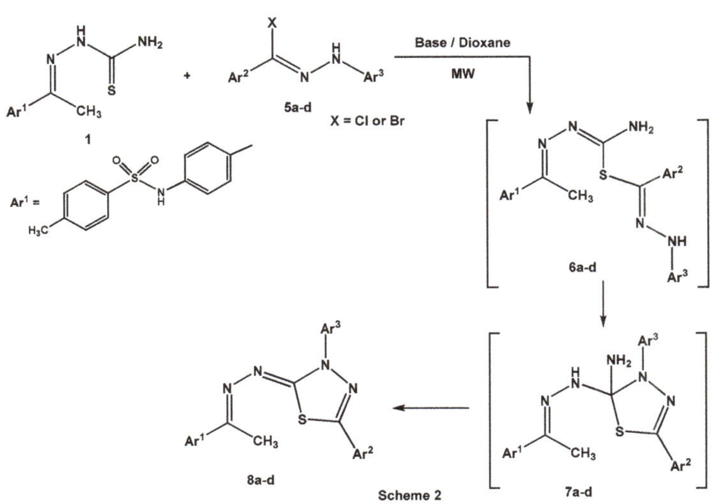

Scheme 2. Synthesis of hydrazono[1,3,4]thiadiazoles **8a–d**.

Table 3. Effect of the nature of basic catalyst on the yield of the products **8a–d**.

Compound Number	Ar^2	Ar^3	Time (min)	Yield (%)	
				Et_3N	Chitosan/MgO Nanocomposite
8a	C_6H_5	C_6H_5	30	74	86
8b	C_6H_5	$4\text{-}NO_2C_6H_4$	35	76	88
8c	$4\text{-}CH_3C_6H_4$	C_6H_5	40	79	85
8d	$4\text{-}CH_3OC_6H_4$	C_6H_5	40	78	87

As shown in Table 3, the percentage yields of the products **8a–d**, the chitosan-MgO nanocomposite as a basic catalyst prevailed over triethylamine under microwave irradiation. The characterization of 2-hydrazono[1,3,4]thiadiazoles **8a–d** was consistent with spectral data (IR, ^{1}H NMR, ^{13}C NMR, and MS (Mass spectroscopy)) and elemental analyses (see Supplementary Materials). As shown in Scheme 2, the reaction proceeded through nucleophilic displacement of the thiol group to the halogen atom to give *S*-alkylated intermediate products **6a–d** [25]. Intramolecular Michael addition [26] of the NH group into the electrophilic carbon atom of (C=N-N=) for intermediates **6a–d** led to the formation of cycloadducts **7a–d**. Elimination of ammonia from the latter intermediates **7a–d** gave the final products **8a–d** (Scheme 2).

4. Conclusions

Recently, nanoparticles (NPs) have been developed as promising candidates in various applications due to their unique properties. In this article, a chitosan-MgO nanocomposite (as a green recyclable biocatalyst) was prepared and well-characterized using FTIR, FESEM, and EDX spectra. The average size of the MgO particles was found to be approximately 6–11 nm for 10 wt% and it was found that the particles' size slightly decreased with increasing magnesia content. This nanocomposite was then used successfully as a heterogeneous basic catalyst for the synthesis of two series of hydrazonothiazoles and hydrazono [1,3,4] thiadiazoles, with sulfonamide moiety, in a comparative study with triethylamine (as a traditional catalyst). In addition to the preferable green impact, the acquired results showed that the chitosan-MgO nanocomposite was a more powerful catalyst in these reactions as compared to triethylamine. The obvious catalytic potency of the chitosan-MgO nanocomposite was attributed to the obtained nanosized MgO and the synergistic effect that is created by the combination of the basic nature of both MgO and chitosan. Moreover, the nanocatalyst could be easily recovered and reused many times without loss in its catalytic activity. Finally, the biopolymer chitosan-metal oxide combination is a promising hybrid nanocomposite that deserves to be explored in many organic transformations.

Supplementary Materials: The following are available online at http://www.mdpi.com/2079-4991/8/11/928/s1.

Author Contributions: S.M.R. suggested the plan, interpreted the results, and wrote the chemistry part of the manuscript; K.D.K. prepared the chitosan-MgO nanocomposite, interpreted the results, performed the experiments, and wrote the other part of the manuscript; A.A. revised the manuscript.

Funding: This research received no external funding.

Conflicts of Interest: The authors declare no conflict of interest.

References

1. Babaie, M.; Sheibani, H. Nanosized magnesium oxide as a highly effective heterogeneous base catalyst for the rapid synthesis of pyranopyrazoles via a tandem four-component reaction. *Arab. J. Chem.* **2011**, *4*, 159–162. [CrossRef]
2. Kumar, D.; Reddy, V.B.; Mishra, B.G.; Rana, R.K.; Nadagouda, M.N.; Varma, R.S. Nanosized magnesium oxide as catalyst for the rapid and green synthesis of substituted 2-amino-2-chromenes. *Tetrahedron* **2007**, *63*, 3093–3097. [CrossRef]

3. Mirzaei, H.; Davoodnia, A. Microwave Assisted Sol-Gel Synthesis of MgO Nanoparticles and Their Catalytic Activity in the Synthesis of Hantzsch 1,4-Dihydropyridines. *Chin. J. Catal.* **2012**, *33*, 1502–1507. [CrossRef]
4. Al-Matar, H.M.; Khalil, K.D.; Meier, H.; Kolshorn, H.; Elnagdi, M.H. Chitosan as heterogeneous catalyst in Michael additions: The reaction of cinnamonitriles with active methylene Moieties and phenols. *Arkivoc* **2008**, 288–301. [CrossRef]
5. Khalil, K.D.; Al-Matar, H.M.; Elnagdi, M.H. Chitosan as eco-friendly heterogeneous catalyst in Michael type additions: Simple and efficient route to pyridones and phthalazines. *Eur. J Chem.* **2010**, *1*, 252–258. [CrossRef]
6. Khalil, K.D.; Al-Matar, H.M. Chitosan based heterogeneous catalyses: 4-Vinylpyridine grafted chitosan as catalyses for Michael additions and alkylpyridazinyl carbonitrile oxidation. *Molecules* **2013**, *18*, 5288–5305. [CrossRef] [PubMed]
7. Pramod, K.S.; Praveen, K.S.; Sushil, K.G.; Dau, D.A. Chitosan: An efficient, reusable, and biodegradable catalyst for green synthesis of heterocycles. *Ind. Eng. Chem. Res.* **2014**, *53*, 2085–2091. [CrossRef]
8. Sanuja, S.; Agalya, A.; Umapathy, M.J. Studies on Magnesium Oxide Reinforced Chitosan Bionanocomposite Incorporated with Clove Oil for Active Food Packaging Application. *Int. J. Polym. Mater. Biopolym. Mater.* **2014**, *63*, 733–740. [CrossRef]
9. Basumallick, S.; Santra, S. Chitosan coated copper-oxide nano particles: A novel electro-catalyst for CO_2 reduction. *RSC Adv.* **2014**, *4*, 63685–63690. [CrossRef]
10. Guibal, E. Heterogeneous catalysis on chitosan-based materials: A review. *Prog. Polym. Sci.* **2005**, *30*, 71–109. [CrossRef]
11. Zablotskaya, A.; Segal, I.; Geronikaki, A.; Eremkina, T.; Belyakov, S.; Petrova, M.; Shestakova, I.; Zvejniece, V.; Nikolajeva, V. Synthesis, physicochemical characterization, cytotoxicity, antimicrobial, anti-inflammatory and psychotropic activity of new *N*-[1,3-(benzo)thiazol-2-yl]-ω-[3,4-dihydroisoquinolin-2(1*H*)-yl]alkanamides. *Eur. J. Med. Chem.* **2013**, *70*, 846–856. [CrossRef] [PubMed]
12. Djukicm, M.; Fesatidou, M.; Xenikakis, I.; Geronikaki, A.; Angelova, V.T.; Savic, V.; Pasic, M.; Krilovic, B.; Djukic, D.; Gobeljic, B.; et al. In vitro antioxidant activity of thiazolidinone derivatives of 1,3-thiazole and 1,3,4-thiadiazole. *Chem. Biol. Interact.* **2018**, *286*, 119–131. [CrossRef] [PubMed]
13. Hu, Y.; Li, C.-Y.; Wang, X.-M.; Yang, Y.-H.; Zhu, H.-L. [1,3,4]Thiadiazole: Synthesis, reactions, and applications in medicinal, agriculture, and materials chemistry. *Chem. Rev.* **2014**, *114*, 5572–5610. [CrossRef] [PubMed]
14. Riyadh, S.M.; El-Motairi, S.A.; Ahmed, H.E.A.; Khalil, K.D.; Habib, E.E. Synthesis, Biological Evaluation, and Molecular Docking of Novel Thiazoles and [1,3,4]Thiadiazoles Incorporating Sulfonamide Group as DHFR Inhibitors. *Chem. Biodiver.* **2018**, *15*, e1800231. [CrossRef] [PubMed]
15. Abbas, E.M.H.; Gomha, S.M.; Farghaly, T.A.; Abdalla, M.M. Synthesis of New Thiazole Derivatives as Antitumor Agents. *Curr. Org. Synth.* **2016**, *13*, 456–465. [CrossRef]
16. Al-Bogami, A.S.; Saleh, T.S.; Mekky, A.E.M.; Shaaban, M.R. Microwave assisted regioselective synthesis and 2D-NMR studies of novel azoles and azoloazines utilizing fluorine-containing building Blocks. *J. Mol. Struct.* **2016**, *1121*, 167–179. [CrossRef]
17. Eweiss, N.F.; Osman, A. Synthesis of Heterocycles Part II. New Routes to Acetylthiadiazolines and Alkylazothiazoles. *J. Heterocycl. Chem.* **1980**, *17*, 1713–1717. [CrossRef]
18. Mosselhi, M.A.N.; Abdallah, M.A.; Mohamed, Y.F.; Shawali, A.S. Synthesis and Tautomeric Structure of 7-Arylhydrazono-7H-[1,2,4]Triazolo[3,4-b][1,3,4]Thiadiazines. *Phosphorous Sulfur Silicon* **2002**, *177*, 487–496. [CrossRef]
19. Laude, B.; Soufiaoui, M.; Arriau, J. Cycloadditions dipolaires-1,3 II. Addition des diarylnitrilimines au *N*-methylindole. Etude experimentale et essai d'interpretation. *J. Heterocycl. Chem.* **1977**, *14*, 1183–1190. [CrossRef]
20. Wolkoff, P. A New Method of Preparing Hydrazonyl Halides. *Can. J. Chem.* **1975**, *53*, 1333–1335. [CrossRef]
21. El Kadib, A.; Primo, A.; Molvinger, K.; Boumina, M.; Brunel, D. Nanosized vanadium, tungsten, and molybdenum oxide clusters grown in porous chitosan microspheres as promising hybrid materials for selective alcohol oxidation. *Chem. A Eur. J.* **2011**, *17*, 7940–7946. [CrossRef] [PubMed]
22. Agrawal, N.R.; Bahekar, S.P.; Agrawal, A.R.; Sarode, P.B.; Chandak, H.S. Cascade Michael-Aldol reaction: Efficient annulation of sulfonamide chalcones into novel cyclohexenones under solvent-free conditions. *Arkivoc* **2016**, 227–245. [CrossRef]

23. De Oliveira Cardoso, M.V.; de Siqueira, L.R.P.; de Silva, E.B.; Costa, L.B.; Hernandes, M.Z.; Rabello, M.M.; Ferreira, R.S.; de Cruz, L.F.; Moreira, D.R.M.; Pereira, V.R.A.; et al. 2-Pyridyl thiazoles as novel anti-Trypanosoma cruzi agents: Structural design, synthesis and pharmacological evaluation. *Eur. J. Med. Chem.* **2014**, *86*, 48–59. [CrossRef] [PubMed]
24. Rostom, S.A.F.; Faidallah, H.M.; Radwan, M.F.; Badr, M.H. Bifunctional ethyl 2-amino-4-methylthiazole-5-carboxylate derivatives: Synthesis and in vitro biological evaluation as antimicrobial and anticancer agents. *Eur. J. Med. Chem.* **2014**, *76*, 170–181. [CrossRef] [PubMed]
25. Geies, A.A.; Kamal-Eldeen, A.M.; Abdelhafez, A.A.; Gaber, A.M. Synthesis of some thiazolo[3,2-a]pyrimidines. *Phosphorous Sulfur Silicon Relat. Elem.* **1991**, *56*, 87–93. [CrossRef]
26. Ranu, B.C.; Banerjee, S. Ionic Liquid as Catalyst and Reaction Medium. The Dramatic Influence of a Task-Specific Ionic Liquid, [bmIm]OH, in Michael Addition of Active Methylene Compounds to Conjugated Ketones, Carboxylic Esters, and Nitriles. *Org. Lett.* **2005**, *7*, 3049–3052. [CrossRef] [PubMed]

nanomaterials

MDPI

Article

Synergistic Effect of Oxygen Vacancies and Ni Species on Tuning Selectivity of Ni/ZrO_2 Catalyst for Hydrogenation of Maleic Anhydride into Succinic Anhydride and γ-Butyrolacetone

Lili Zhao [1], Jianghong Zhao [1], Tianjie Wu [1], Min Zhao [2], Wenjun Yan [2], Yin Zhang [1], Haitao Li [1], Yongzhao Wang [1], Tiancun Xiao [3,*] and Yongxiang Zhao [1,*]

1 Engineering Research Center of Ministry of Education for Fine Chemicals, School of Chemistry and Chemical Engineering, Shanxi University, Taiyuan 030006, China; lzhao@sxu.edu.cn (L.Z.); zhaojianghong@sxu.edu.cn (J.Z.); 201722907009@email.sxu.edu.cn (T.W.); sxuzhy@sxu.edu.cn (Y.Z.); htli@sxu.edu.cn (H.L.); catalyst@sxu.edu.cn (Y.W.)

2 Institute of Coal Chemistry, Chinese Academy of Sciences, Taiyuan 030001, China; zhaomin@sxicc.ac.cn (M.Z.); yanwenjun@sxicc.ac.cn (W.Y.)

3 Inorganic Chemistry Laboratory, Oxford University, Oxford, OX1 3QR, UK

* Correspondence: xiao.tiancun@chem.ox.ac.uk (T.X.); yxzhao@sxu.edu.cn (Y.Z.)

Received: 26 January 2019; Accepted: 5 March 2019; Published: 11 March 2019

Abstract: ZrO_2 nanoparticles, ZrO_2 (P) and ZrO_2 (H), with different tetragonal phase contents, were prepared. ZrO_2 (P) possessed higher tetragonal phase content than ZrO_2 (H). Ni/ZrO_2 catalysts (10% (w/w)), using ZrO_2 (P) and ZrO_2 (H) as supports, were prepared using an impregnation method, and were characterized using XRD, Raman, H_2-TPR, XPS, and H_2-TPD techniques. Their catalytic performance in maleic anhydride hydrogenation was tested. The Ni/ZrO_2 (P) catalyst exhibited stronger metal-support interactions than the Ni/ZrO_2 (H) catalyst because of its higher number of oxygen vacancies and the low-coordinated oxygen ions on its surface. Consequently, smaller Ni crystallites and a higher C=C hydrogenation activity for maleic anhydride to succinic anhydride were obtained over a Ni/ZrO_2 (P) catalyst. However, the C=O hydrogenation activity of Ni/ZrO_2 (P) catalyst was much lower than that of the Ni/ZrO_2 (H) catalyst. A 43.5% yield of γ-butyrolacetone was obtained over the Ni/ZrO_2 (H) catalyst at 210 °C and 5 MPa of H_2 pressure, while the yield of γ-butyrolactone was only 2.8% over the Ni/ZrO_2 (P) catalyst under the same reaction conditions. In situ FT-IR characterization demonstrated that the high C=O hydrogenation activity for the Ni/ZrO_2 (H) catalyst could be attributed to the surface synergy between active metallic nickel species and relatively electron-deficient oxygen vacancies.

Keywords: maleic anhydride; oxygen vacancies; selective hydrogenation; Ni/ZrO_2

1. Introduction

Maleic anhydride (MA), as the third most important anhydride in commercial use, can be hydrogenated to produce succinic anhydride (SA), γ-butyrolacetone (GBL), 1,4-butanediol (BDO), and tetrahydrofuran (THF) products (Figure 1). SA is an important raw material of biodegradable plastic polybutylene succinate (PBS), and GBL is currently one of the most valuable and environmentally friendly media [1,2]. Hence, much attention has been focused on the selective hydrogenation of MA to SA or GBL. However, the process remains a challenge because of the coupled structure of the C=C and C=O bonds in MA molecules [3]. The coupled molecular structure leads to a delocalization of the electron density in the C=C and C=O bonds. This makes it difficult for the selectively hydrogenation of the C=C bond to obtain SA, or for the C=C and C=O bonds to obtain GBL. In addition, MA has a different molecular structure from other linear conjugated molecules, such as crotonaldehyde or acrolein, in that it is a compound with a five-membered cyclic structure. The special geometric structure of the reactant molecule will affect its adsorption mode on catalysts and the corresponding hydrogenation mechanism [4]. Therefore, the tailoring of high-activity and high-selectivity catalysts to obtain SA or GBL is an important but challenging subject.

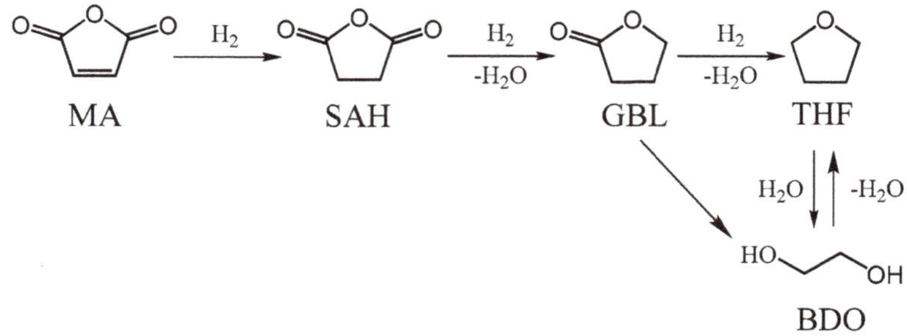

Figure 1. Reaction scheme of hydrogenation of maleic anhydride (MA).

In the current literature, the metal catalysts Pd, Ru, Cu, and Ni are widely accepted as being active in the hydrogenation of MA. Among them, the Ni catalyst has attracted a great deal of attention because of its high hydrogenation activity and low cost [5–9]. However, because of their weak hydrogenation ability towards the C=O bond of Ni [10,11], hydrogenation products are mostly mixtures of SA and GBL. Some valuable strategies have been employed to regulate the selectivity of Ni-based catalysts [12–16], such as introducing a second component as a promoter and modifying surface acid-base properties of supports. Results of experiments have also shown that these are effective regulation methods. In addition to these measures, the product distribution in MA hydrogenation could also be modulated by using reducible oxides as supports (e.g., CeO_2 and TiO_2) [17–20]. Our research group found that the Ni/CeO_2 catalyst exhibited higher C=O hydrogenation activity when compared with the Ni/Al_2O_3 catalyst, and the superior catalytic activity in C=O hydrogenation was ascribed to the reduction of CeO_2 [21]. In the hydrogenation of other carbonyl compounds and CO_2 methanation reactions, it has been further found that the catalysts with abundant oxygen vacancies on the surface, such as CeO_2, TiO_2 and Co_3O_4 supported metal catalysts, exhibited superior catalytic performance in hydrodeoxygenation. The researchers concluded that the superior performances of these catalysts stemmed from the promotion of oxygen vacancies [22–26].

Recently, ZrO_2 has received considerable attention and has been applied in a variety of reactions because of its amphoteric properties and the multiformity of its crystalline phases [27,28]. The morphology of ZrO_2 was found to play a vital role in various catalytic reactions. Samson et al. found that when ZrO_2 was present in the tetragonal phase, it showed a higher activity towards methanol synthesis from CO_2 [29], whereas Rhodes et al. found that monoclinic ZrO_2 was more active for methanol synthesis [30]. Amorphous ZrO_2 was found to be beneficial for the dry reforming of the methane reaction [31]. The aforementioned experimental results, concerning the effects of the ZrO_2 crystalline phase, seem to be contradictory. Whether the crystalline phase of ZrO_2 is directly related to the catalytic activity of ZrO_2 supported metal catalysts is unknown. Inspired by the reducible oxide-supported metal catalytic system, we speculate that the reason for the different catalytic behaviors of ZrO_2-based catalysts might be the different surface structures of catalysts. Compared to CeO_2 and TiO_2, ZrO_2 is more difficult to reduce. Whether oxygen vacancies are formed on the Ni/ZrO_2 catalyst, and whether those oxygen vacancies affect catalytic performance in MA hydrogenation, until now, remains unclear.

Therefore, in an effort to better understand the formation of oxygen vacancies on Ni/ZrO_2 catalysts and the effect of oxygen vacancies on MA hydrogenation, we prepared two Ni/ZrO_2 catalysts with different surface structures and tested their catalytic performance in MA hydrogenation. The results are presented in this paper.

2. Experimental Section

2.1. ZrO_2 Preparation

ZrO_2 (P) was prepared as follows [32]: firstly, a white precipitate was obtained by refluxing a mixture of a 0.5 M solution of zirconyl nitrate ($ZrO(NO_3)_2 \cdot 2H_2O$; >45.0% ZrO_2, Beijing Chemicals, Beijing, China) at 100 °C for 240 h; during the process the solution pH value maintained at 1.5 by dropwise addition of ammonia solution. Then, the obtained precipitate was transferred into a Teflon-lined, stainless-steel autoclave (100 mL) and heated in an oven at 110 °C for 4 h. The final precipitate was washed with absolute alcohol until pH = 7, and then dried at 110 °C for 12 h before being calcined at 400 °C for 4 h.

ZrO_2 (H) was prepared using a hydrothermal method [33]. It was synthesized at 140 °C under autogenous pressure for 2.5 h in a Teflon-lined stainless-steel autoclave (100 mL), which contained solutions (80 mL) of urea ($CO(NH_2)_2$; >99.9%, Beijing Chemicals, Beijing, China) and zirconyl nitrate ($ZrO(NO_3)_2 \cdot 2H_2O$; >45.0%, Beijing Chemicals, Beijing, China). The concentration of Zr^{4+} in the solutions was 0.2 M, and the urea/Zr^{4+} molar ratio was 10. The resulting precipitates were washed with absolute alcohol until they reached pH = 7, dried at 110 °C for 12 h, and then calcined at 400 °C for 4 h.

2.2. Ni/ZrO_2 Preparation

NiO/ZrO_2 catalysts were prepared using the impregnation method. Typically, 1.0 g of ZrO_2 was added to an aqueous solution, consisting of 0.5476 g of nickel nitrate and 2.2 mL H_2O, under vigorous stirring, after which the sample was dried at 120 °C for 12 h, and then calcined in air at 450 °C for 3 h. After the calcination treatment, the sample was denoted as NiO/ZrO_2. Following that, the samples were reduced at 400 °C for 3 h in an H_2 flow (30 mL/min), denoted as Ni/ZrO_2.

2.3. Structure Characterizations

The nickel content in the Ni/ZrO_2 catalysts was determined using inductively coupled plasma (ICP) on an iCAP 7400 ICP-OES (Thermo Fisher Scientific, Waltham, MA, USA).

The specific surface areas of the ZrO_2 and ZrO_2 supported catalysts were measured by N_2 physisorption, at −196 °C, and using an ASAP-2020 instrument (Micromeritics, Atlanta, GA, USA).

X-ray diffraction (XRD) of the samples (ZrO_2 support, NiO/ZrO_2 and Ni/ZrO_2) was performed using an X-ray diffractometer (Bruker D8 Advance, Karlsruhe, Germany) with Cu Kα radiation (λ = 1.54056 Å). The operating voltage and current were 40 kV and 40 mA, respectively.

Raman spectra were obtained using a Lab RAM HR Evolution Raman microscope (Horiba Scientific, Paris, France). The visible and UV Raman spectra were obtained using Ar^+ (532 nm) and He-Cd lasers (325 nm) as excitation sources, respectively. A quantitative determination of the tetragonal phase x(T) content, present in each sample, was estimated using the following equation [34]:

$$X(T) = I(T)/[I(T) + I(M)]$$

where I(T) represents the added intensities of the two bands at ~148 and 269 cm^{-1}, which are characteristic of the tetragonal phase, and I(M) denotes the added intensities of the two bands at 178 and 191 cm^{-1}, and which are associated with the monoclinic phase.

The high-resolution transmission electron microscopy (HRTEM) images were measured on a JEOL JEM-2010 (Tokyo, Japan), which operated at 200 kV. Before taking the measurements, the NiO/ZrO_2 samples were reduced at 400 °C for 3 h in H_2 flow (30 mL/min) and then cooled to room temperature. Following that, the samples were transferred to a beaker containing anhydrous ethanol under N_2 protection. Then, the samples were ultrasonically dispersed in ethanol and supported on a carbon-coated copper grid. High-angle annular dark-field scanning transmission electron microscopy

(HAADF-STEM) and energy dispersive spectrometer (EDS) mapping images of the samples were obtained using a JEOL JEM-2010F (Tokyo, Japan) at 200 kV.

X-ray photoelectron spectroscopy (XPS) measurements were taken using a Kratos AXIS Ultra DLD spectrometer (Manchester, UK) with a monochromatic Al Kα (1486.6 eV) irradiation source. The X-ray gun operated at 150 W. The survey spectra were recorded with a pass energy of 160 eV, and the high-resolution spectra were recorded with a pass energy of 40 eV. The sampling area was 300×700 μm^2. The binding energy was corrected by setting the C1s peak to 284.6 eV. For the ZrO_2 samples' test, they were placed into an XPS sample cell, which was then pumped down to 10^{-8} Pa before the spectra were recorded. For the test of the Ni/ZrO_2 sample, the NiO/ZrO_2 samples were first placed into an XPS sample cell, reduced at 400 °C for 3 h, and then cooled down to room temperature in H_2 flow (30 mL/min). Subsequently, the sample cell was pumped down to 10^{-8} Pa, and then the spectra were recorded.

H_2 temperature-programmed reduction (H_2-TPR) was performed on a Micromeritics Auto Chem II 2920 (Atlanta, GA, USA) equipped with a thermal conductivity detector to determine the reducibility of the catalysts. First, 30 mg of Ni/ZrO_2 sample were treated in Ar at 300 °C for 1 h and then cooled to 50 °C. Subsequently, the H_2-TPR profiles were recorded while heating the samples in H_2/Ar (10% *v*/*v*) with 50 mL/min of gas flow, from 50–700 °C at a ramp of 10 °C/min.

H_2 temperature-programmed desorption (H_2-TPD) measurements were carried out on the same apparatus as was used for the H_2-TPR (Micromeritics Auto Chem II 2920, Atlanta, GA, USA). First, a 100 mg NiO/ZrO_2 sample was first in situ reduced at 400 °C for 3 h in pure H_2 and then cooled down to 50 °C. It was then purged with Ar for 1 h at 50 °C to remove the excess hydrogen adsorbed on the surface. H_2/Ar (10% *v*/*v*) was then injected at 50 mL/min until saturation. Ar was used to flush the sample until the baseline was stable. H_2-TPD profiles were recorded up to 700 °C at a heating rate of 10 °C/min.

In-situ FT-IR spectra of cyclohexanone were collected on a spectrometer (Bruker Tensor 27, Karlsruhe, Germany). 0.02 g NiO/ZrO_2 sample were placed into an IR cell. Prior to the adsorption of cyclohexanone, the sample was reduced at 400 °C for 3 h in H_2 flow (30mL/min) and then cooled to 210 °C. Following that, the IR cell containing the samples was pumped down to $<6 \times 10^{-3}$ Pa and a spectrum was recorded as the background. Gas cyclohexanone molecules were then introduced to the IR cell for the adsorption for 60 min. It was then desorbed, via vacuum pumping down to $<6 \times 10^{-3}$ Pa. The spectra were recorded with a resolution of 2 cm^{-1}.

2.4. Catalytic Activity Tests

The catalytic performances of the Ni/ZrO_2 catalysts in the hydrogenation of MA were measured in a batch reactor (100 mL) with mechanical agitation at 210 °C and 5 MPa of H_2 pressure. Before the test, the catalysts were pre-reduced using a stream of H_2 (30 mL/min) in a quartz tube at 400 °C for 3 h and then cooled down to room temperature. Meanwhile, the MA (4.9 g) and THF (the purity of THF $\geq$ 99.99%, $H_2O \leq 20$ ppm) were charged in the autoclave. Then, the reduced catalyst (0.1 g) (40–60 mesh) was charged in the autoclave under N_2 protection. Before each run, the autoclave was sealed and flushed with N_2 three times and H_2 five times to achieve a system pressure of 5 MPa. The reactor was heated to 210 °C, and the agitator operated at 400 rpm.

Different solvent such as 1,4-dioxane, cyclohexane were investigated. The results showed that the hydrogenation products were SA and GBL, with no THF and other products being detected. The carbon balance was between 95 and 105%. When THF was used as solvent, products SA and GBL are detected, and the carbon balance calculated based on the sum of SA and GBLwas between 95 and 105%. This suggest that there were no deep hydrogenation products like THF, BDO produced using THF as solvent for the present Ni/ZrO_2 catalysts system.

The samples obtained from the reactor were analyzed using a gas chromatograph (Agilent, 7890B, Palo Alto, CA, USA) equipped with a DB-5 capillary column and FID detector. The conversion of MA and the selectivity to the product were calculated using the following equations:

$$Conversion\ (MA) = (MA_{in} - MA_{out})/MA_{in} \times 100\%$$

$$Selectivity\ (i) = Product_{i,\ out}/\sum product_{i,\ out} \times 100\%$$

where MA_{in}, MA_{out} and $Product_{i,out}$ represent the molar concentration of the inlet reactant, outlet reactant, and outlet product of species i, respectively.

3. Results

3.1. X-Ray Powder Diffraction (XRD) Patterns

The crystalline structures of the ZrO_2 supports and the corresponding supported nickel catalysts were examined using XRD (Figure 2). This showed that the ZrO_2 (P) and ZrO_2 (H) possessed the characteristic lines of a mixture of monoclinic (JCPDS 65-1023) and tetragonal (JCPDS 81–1544) zirconia. The Rietveld method was used for diffraction peak deconvolution, and the calculated content of each phase is listed in Table 1 [35]. The results showed that there was a higher tetragonal phase (t-ZrO_2) content for ZrO_2 (P) than that for ZrO_2 (H). The crystallite size of the ZrO_2 (P) and the ZrO_2 (H) calculated using the Scherrer equation are 16 and 10 nm, respectively. Based on "nanoparticle size effect" [36], the tetragonal phase can be stabilized at room temperature below a critical size (30 nm), which is due to the generation of excess oxygen vacancies; therefore, it could be supposed that more oxygen vacancies existed in ZrO_2 (P) than in ZrO_2 (H).

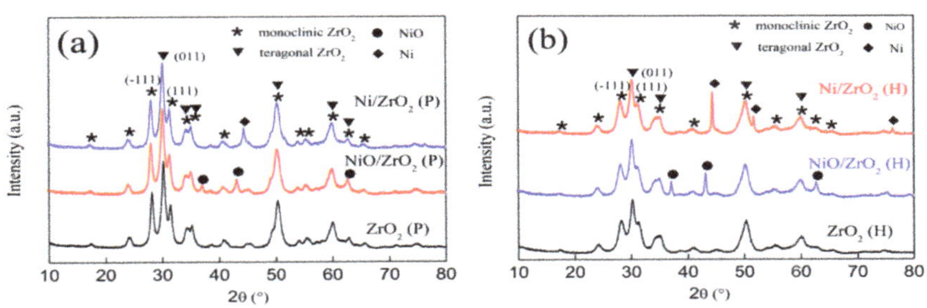

Figure 2. X-ray powder diffraction patterns of (**a**) ZrO_2 (P) support and supported Ni catalysts and (**b**) ZrO_2 (H) support and supported Ni catalysts.

Figure 2 showed that new peaks appeared in the diffraction patterns of the NiO/ZrO_2 (P) and NiO/ZrO_2 (H) samples, which were attributed to crystalline NiO species (JCPDS 22-1189). The calculated mean crystallite size of NiO in the NiO/ZrO_2 (P) was approximately 16 nm, which was smaller than the 26 nm of the NiO/ZrO_2 (H) (Table 1). For the reduced Ni/ZrO_2 (P) and Ni/ZrO_2 (H) catalysts, the NiO characteristic diffraction peaks disappeared accompanied by the appearance of Ni characteristic peaks. The calculated mean crystallite size of Ni in the Ni/ZrO_2 (P) was approximately 18 nm, while that of Ni/ZrO_2 (H) grew to 40 nm. The crystallite size of ZrO_2 was unchanged, even after calcination and reduction treatment for either the Ni/ZrO_2 (P) or Ni/ZrO_2 (H) catalysts.

Table 1. A summary of metal content, specific surface area and pore diameter, calculated tetragonal phase content, and crystalline size of the ZrO_2 (P) and ZrO_2 (H) supports and their corresponding supported nickel catalysts.

Catalysts	Metal Content [a] [wt %]	S_{BET} [m^2g^{-1}]	Pore Diameter [nm]	Tetragonal Phase [b] [%]	Tetragonal Phase [c] [%]	Tetragonal Phase [d] [%]	Crystalline Size of ZrO_2 [e] [nm]	Crystalline Size of NiO [f] [nm]	Crystalline Size of Ni [g] [nm]
ZrO_2 (P)	-	98	7.1	48	distorted	23	16	-	-
NiO/ZrO_2 (P)	-	73	5.8	47	-	disordered	17	16	-
Ni/ZrO_2 (P)	9.42	53	3.5	46	-	-	17	-	18
ZrO_2 (H)	-	71	3.8	39	35	11	10	-	-
NiO/ZrO_2 (H)	-	58	3.7	41	-	distorted	10	26	-
Ni/ZrO_2 (H)	9.39	25	3.6	41	-	-	10	-	40

[a] Measured by inductively coupled plasma (ICP). [b] Calculated using the Rietveld method from the X-ray powder diffraction data. Calculated using the Raman data at a [c] 325 nm and [d] 532 nm excitation wavelengths. Calculated using the Scherrer equation from the[e] (-111) plane of m-ZrO_2, [f] (012) plane of NiO, and [g] the (111) plane of Ni.

3.2. Raman Spectra

Raman spectroscopy was used to further detect the microstructures of the samples due to its sensitivity to oxygen displacement and intermediate range order of the samples [37]. Excitation wavelengths of 325 and 532 nm were used to detect the phases, from the surface to the deeper inner part of the catalyst, which resulted from light absorption and light scattering $\{I \propto (1/\lambda)^4\}$ [38]. When excited by the 532 nm laser (Figure 3a), both the ZrO_2 (H) and ZrO_2 (P) exhibited intense bands of Ag at 178 and 191 cm^{-1}, Bg at 222, 333, and 382 cm^{-1}, Ag at 476 cm^{-1}, and Bg at 615 cm^{-1}, which were assigned to the monoclinic ZrO_2 (m-ZrO_2) and the band Eg at 269 cm^{-1}, which was ascribed to tetragonal ZrO_2 (t-ZrO_2) [39].

The quantitative determination of the tetragonal phase content of each sample is shown in Table 1. The content of tetragonal phase for ZrO_2 (P) was 23%—higher than the 11% of ZrO_2 (H)—showing that more oxygen vacancies existed in ZrO_2 (P) than in ZrO_2 (H). With 325 nm laser excitation (Figure 3b), both the ZrO_2 (H) and the ZrO_2 (P) revealed a monoclinic ZrO_2 stretching peak. The difference is that the band of Eg was at 269 cm^{-1} for ZrO_2 (H), while a band centered at 256 cm^{-1} appeared for ZrO_2 (P). The peak, centered at 269 cm^{-1}, was a typical characteristic peak for the tetragonal phase and showed the characteristics of an asymmetric Zr-O-Zr stretching mode [40]. It was reported that the shift of this peak to a lower wavenumber is due to the movement of oxygen [41]. The calculated results of the XRD and the 532 nm Raman spectroscopy showed that the tetragonal phase content in the ZrO_2 (P) was higher than that in the ZrO_2 (H). Thus, the peak at 256 cm^{-1} is ascribed to a decrease in the symmetry of the tetragonal phase structure, which was caused by the higher number of oxygen vacancies.

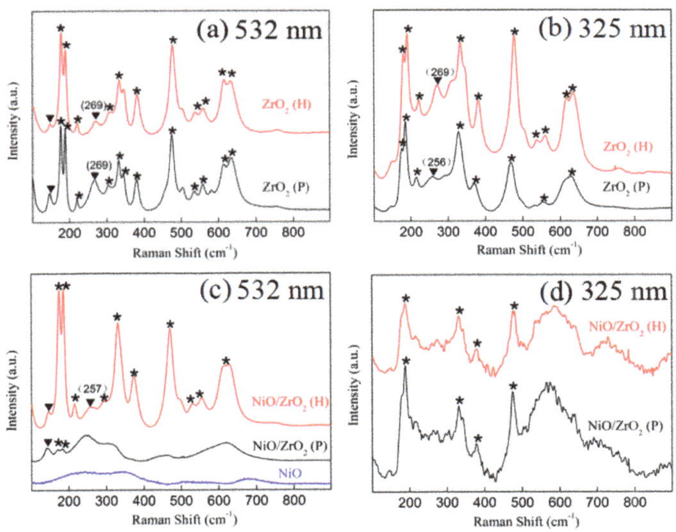

Figure 3. Raman spectra of ZrO_2 supports and NiO/ZrO_2 catalysts excited at 532 and 325 nm. ★ denotes monoclinic ZrO_2; ▼ denotes tetragonal ZrO_2. (**a**) Raman spectra of ZrO_2 supports excited at 532 nm, (**b**) Raman spectra of ZrO_2 supports excited at 325 nm, (**c**) Raman spectra of NiO/ZrO_2 samples excited at 532 nm, (**d**) Raman spectra of NiO/ZrO_2 samples excited at 325 nm.

As shown in Figure 3d, the NiO/ZrO_2 (H) and NiO/ZrO_2 (P) catalysts exhibited typical monoclinic phase behavior with no discernible differences at 325 nm laser excitation. Upon being excited by a 532 nm laser, stronger and more well-defined Raman peaks were obtained (Figure 3c), which gave accurate overall structure information of the NiO/ZrO_2 samples. For the NiO/ZrO_2 (H), the Raman spectrum is dominated by strong bands which were attributed to m-ZrO_2 and a less prominent broad band at 257 cm^{-1}. The slight shift of the band from 269 cm^{-1} for ZrO_2 (H) to 257 cm^{-1}

for NiO/ZrO_2 (H) could be ascribed to the increase in the number of oxygen vacancies. This was most likely caused by the interaction between nickel species and ZrO_2 (H). The NiO/ZrO_2 (P) spectrum showed only broad continuum lines with poorly defined bands maxima at 245, 448 and 620 cm^{-1}. which were attributed to a breakdown of the wave-vector selection rule by translational disorder caused by the random substitution of vacancies or cations [42]. This meant that the interaction between the nickel species and ZrO_2 (P) support was stronger, which led to a disordered tetragonal structure.

A comparison of the XRD and Raman results showed no discernible change in the ZrO_2 structure after the nickel loading in the XRD, whereas the Raman spectra exhibited significant changes in the tetragonal structure for the NiO/ZrO_2 (P) and NiO/ZrO_2 (H) catalysts, which also provides strong evidence that Raman is more sensitive to the intermediate range structures, while XRD characterizes the long-range ordering of the structures.

3.3. High-Resolution Transmission Electron Microscopy (HRTEM) Images

HRTEM images of the Ni/ZrO_2 catalysts are shown in Figure 4. Metallic Ni with lattice spacings of 0.17 and 0.20 nm were observed for both the Ni/ZrO_2 (P) and Ni/ZrO_2 (H) catalysts. It is noticeable that the interface between the Ni nanoparticles and the ZrO_2 (P) substrate was a coalesced heterostructure, as shown in Figure 4a (and enlarged in Figure 4b). The Ni nanoparticles are embedded into a large ZrO_2 (P) substrate, with an irregular borderline emerging in the disordered interface region. The lattice spacing of the t-ZrO_2 (101) increased from 0.29 to 0.30 nm, indicating that some of the nickel ions incorporated into the t-ZrO_2 lattice through Ni^{2+} dissolution into ZrO_2 (P) or were located at the interstitial sites, resulting in the formation of a Ni-O-Zr structure, and thereby causing lattice expansion [43]. This shows that there was a strong interaction between the closely contacted nickel species and the t-ZrO_2 for the Ni/ZrO_2 (P), which agrees with the Raman results. However, unlike the Ni/ZrO_2 (P), there were large Ni particles surrounded by small ZrO_2 (H) particles for the Ni/ZrO_2 (H) catalyst without the appearance of coalesced structures, as shown in Figure 4d, but with a loosely contacted region at the interfaces (Figure 4e). As shown in Figure 4c,f, smaller nickel particles were uniformly dispersed on the ZrO_2 substrate for the Ni/ZrO_2 (P) catalyst, while the nickel particles were aggregated around the ZrO_2 particles for the Ni/ZrO_2 (H) catalyst.

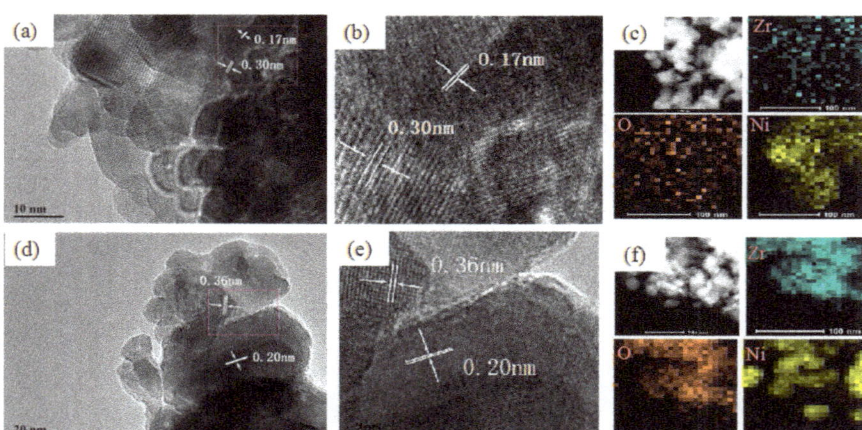

Figure 4. High-resolution transmission electron microscopy (HRTEM) images of (**a**) Ni/ZrO_2 (P) and (**d**) Ni/ZrO_2 (H). Enlarged selected area HRTEM images of (**b**) Ni/ZrO_2 (P) and (**e**) Ni/ZrO_2 (H). HAADF-STEM images and corresponding EDS elemental mapping images of (**c**) Ni/ZrO_2 (P) and (**f**) Ni/ZrO_2 (H).

3.4. H_2 Temperature-Programmed Reduction (H_2-TPR)

The H_2-TPR profiles of ZrO_2 (P), ZrO_2 (H), and their corresponding supported nickel catalysts are presented in Figure 5. No reduction peaks were observed for ZrO_2 (H), while ZrO_2 (P) exhibited an obvious reduction peak at 540 °C, indicating that ZrO_2 (P) was more easily reduced. It was reported that surficial O atoms at low-coordinated sites are easily removed, and that oxygen vacancies facilitate the activation and transportation of active oxygen species, thereby promoting the reducibility of ZrO_2 [44]. From this perspective, the reducibility of ZrO_2 (P) originated from its special surface structure, more low-coordinated oxygen ions and oxygen vacancies on the surface of ZrO_2 (P). Furthermore, these low-coordinated sites introduce defective states in the band gap and enhance the interaction with the deposited metal catalysts [44,45]. This is consistent with our Raman and HRTEM experiment results—i.e., nickel species have a stronger interaction with the ZrO_2 (P) support. Furthermore, it can be understood that the stronger interaction between nickel species and ZrO_2 (P) originates from the abundance of low-coordinated oxygen ions and oxygen vacancies on the surface of ZrO_2 (P).

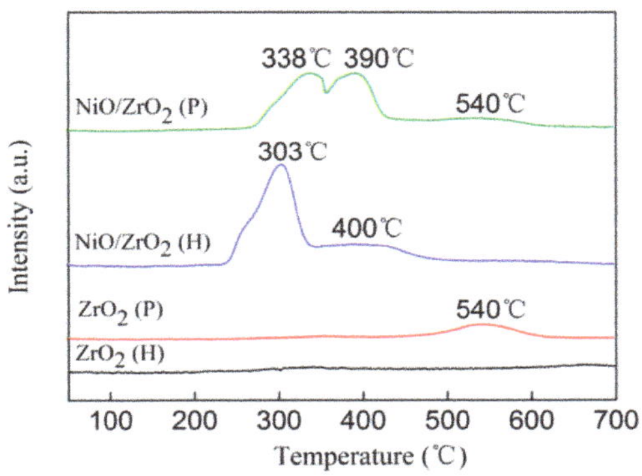

Figure 5. H_2-TPR profiles of ZrO_2 supports and NiO/ZrO_2 catalysts.

The NiO/ZrO_2 (H) showed a sharp reduction peak centered at 303 °C with a minor shoulder peak at 400 °C. The H_2 uptake peak at 303 °C was attributed to the reduction of large NiO that had weak interactions with the support [46]. For the NiO/ZrO_2 (P) sample, three H_2 uptake peaks were observed at 338, 390, and 540 °C. The first peak was assigned to the reduction of NiO particles with weak interactions with the support. The second peak was related to the NiO exhibiting a relatively strong interaction with the support. Compared to the profile of the ZrO_2 (P), the reduction peak at 540 °C could be attributed to the reduction of ZrO_2 [46]. The reduction temperature of the NiO/ZrO_2 (P) was much higher than that of the NiO/ZrO_2 (H), indicating the presence of strong interactions between the nickel species and the ZrO_2 (P). The strong interaction between the nickel species and support could hinder the migration of nickel species during the calcination and reduction procedure. In this way, smaller Ni particles could be obtained in the NiO/ZrO_2 (P) sample. These findings were in in line with the results of XRD and HRTEM.

3.5. XPS Characterization

To explore the properties of the oxygen vacancies on the ZrO_2 support and their changes after loading nickel, XPS was conducted. It is widely accepted that a neutral O vacancy introduces two extra electrons in the lattice, which can be localized either in the created vacancy or in nearby cation sites. In ZrO_2, the extra charge is trapped in the vacancy site rather than reducing the nearest Zr ions. Hence, there were three favored charge states for oxygen vacancies existed on surface of the ZrO_2: a neutral oxygen vacancy with the two electrons remaining at the oxygen vacancy, a singly charged oxygen vacancy, and a doubly charged oxygen vacancy [44]. Due to the decrease in charge density, an increase in the O1s binding energy for the singly-charged oxygen and the doubly-charged oxygen vacancies is inevitable, while the neutral oxygen vacancy peak likely lies at or near the same position as the lattice oxygen ion peak. In K. T. Leung's work on XPS fitting [47], oxygen vacancies were fitted into two types of oxygen vacancies, and the changes in the oxygen vacancies' electronic properties were expounded more clearly. In line with this earlier research, the O1's spectra from the samples in this study were analyzed, using curve fitting and four peaks were assigned to the lattice oxygen (O′), singly charged oxygen vacancies (O″), doubly charged oxygen vacancies (O‴), and the hydroxyl or/and carbonates groups (O⁗) on ZrO_2 (Figure 6).

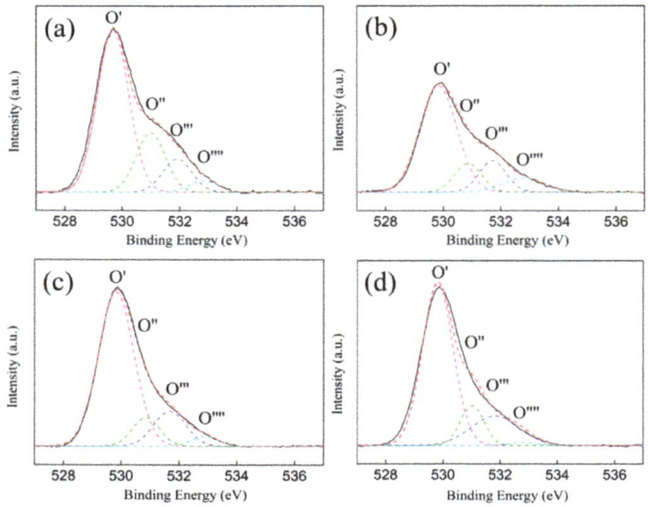

Figure 6. O 1s XPS spectra of the ZrO_2 supports and Ni/ZrO_2 catalysts: (**a**) ZrO_2 (P), (**b**) ZrO_2(H), (**c**) Ni/ZrO_2 (P), and (**d**) Ni/ZrO_2 (H).

Table 2 shows summary details for the binding energy and the surface atomic concentration that was calculated by integrating different oxygen species' peak areas. Two forms of oxygen vacancies were centered at 531.0 (BE_2) and 531.9 eV (BE_3), which corresponded to the singly and doubly charged oxygen vacancies on the ZrO_2 (P), respectively. For the ZrO_2 (H), the binding energies of the singly and doubly charged oxygen vacancies were centered at 530.8 (BE_2) and 531.7 eV (BE_3), respectively. The different binding energies, shown for the same types of oxygen vacancies on the ZrO_2 (P) and the ZrO_2 (H), illustrate that the electron properties of the oxygen vacancies on the ZrO_2 (P) and the ZrO_2 (H) are different. The oxygen vacancy concentration of the ZrO_2 (P) was 33.1%—much higher than that of the ZrO_2 (H) (24.5%). This was consistent with the XRD and Raman results. A comprehensive analysis of the above characterization results shows that the ZrO_2 support, with a higher concentration of electron-deficient oxygen vacancies and low coordination oxygen ion on the surface, demonstrates much stronger interactions with the Ni species.

Table 2. The binding energy of O 1s lines and the corresponding surface atomic concentration of ZrO_2 supports and Ni/ZrO_2 catalysts.

Samples	Binding Energy (eV)				*I* (%)				
	BE_1	BE_2	BE_3	BE_4	I_1 (%)	I_2 (%)	I_3 (%)	I_4 (%)	$I_2 + I_3$ (%)
ZrO_2 (P)	529.7	531.0	531.9	532.8	62.9	21.2	11.9	4.0	33.1
ZrO_2 (H)	529.8	530.8	531.7	532.8	71.0	11.6	12.9	4.5	24.5
Ni/ZrO_2 (P)	529.8	530.8	531.7	532.8	71.8	10.1	14.6	3.5	24.7
Ni/ZrO_2 (H)	529.8	531.0	531.9	532.8	70.5	13.4	13.0	3.1	26.4

After nickel loading, the binding energy of O″ for the Ni/ZrO_2 (P) shifted from 531.0 to 530.8 eV, and the binding energy of O‴ shifted from 531.9 to 531.7 eV. Meanwhile the total oxygen vacancy concentration decreased significantly, from 33.1% to 24.7%. This was potentially because a small amount of the nickel species entered into the oxygen vacancies, and the oxygen vacancies bore extra charges for the charge balance in the Ni-O-Zr like structure [48]. However, the binding energy of O″ and O‴ increased from 530.8 and 531.7 eV to 531.0 and 531.9 eV, respectively, after ZrO_2 (H) loading nickel. Additionally, the total oxygen vacancy concentration increased slightly when compared to that of the ZrO_2 (H) support. This was because the Ni^0 particles, which were first reduced, promoted the generation of additional oxygen vacancies at the Ni/ZrO_2 interface and caused local structural deformation around the vacancy, which has been observed in Raman characterization [49]. Furthermore, the oxygen vacancies that were promoted by the Ni^0 exhibited lower charge densities, and were different from the inherent oxygen vacancies of the ZrO_2 (H).

Figure 7 shows the Ni 2p XPS spectra of the Ni/ZrO_2 catalyst. Three different chemical states of nickel were found in the Ni $2p_{3/2}$ XPS spectra. In the Ni $2p_{3/2}$ XPS spectra, the binding energy of the Ni^0 was situated at 852.2 eV [50], and the relative amount of Ni^0 species on the Ni/ZrO_2 (P) was 57.8%, which was almost equal to that on the Ni/ZrO_2 (H) (58.4%) (Table 3). Two other peaks appeared for the Ni/ZrO_2 (P): one at 853.9 eV was attributed to Ni^{2+}, in the form of NiO, and the other, at 855.6 eV, was attributed to Ni^{2+}, in the form of oxide and hydroxide phases [50]. For the Ni/ZrO_2 (H), the corresponding binding energy of the above two peaks shifted to 853.0 and 855.1 eV, respectively. The higher binding energy of the nickel species in the Ni/ZrO_2 (P) further confirmed that nickel species have a stronger interaction with ZrO_2 (P).

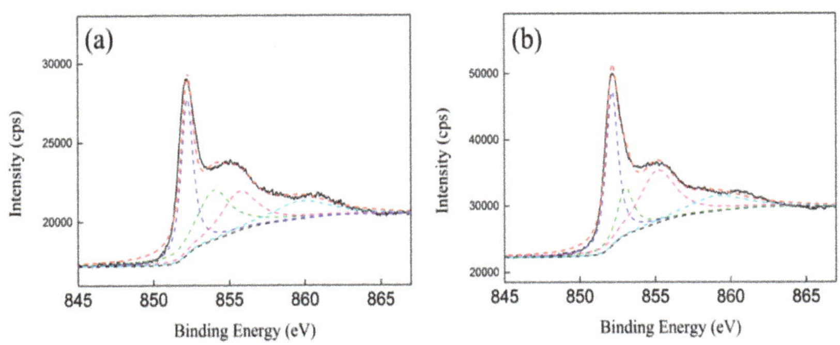

Figure 7. Ni 2p XPS spectra of Ni/ZrO_2 catalysts: (**a**) Ni/ZrO_2 (P) and (**b**) Ni/ZrO_2 (H).

Table 3. Binding energy of Ni 2p lines and the corresponding surface concentration of Ni/ZrO_2 catalysts.

Samples	Ni^0 ($Ni\ 2P_{3/2}$)	NiO ($Ni\ 2P_{3/2}$)	Ni^{2+} ($Ni\ 2P_{3/2}$)	$C(Ni^0)$ (%)	C(NiO) (%)	$C(Ni^{2+})$ (%)
Ni/ZrO_2 (P)	852.2	853.9	855.6	57.8	22.5	19.7
Ni/ZrO_2 (H)	852.2	853.0	855.1	58.4	19.2	22.4

3.6. Catalytic Performances of Ni/ZrO_2 (P) and Ni/ZrO_2 (H) Catalysts

The hydrogenation of MA over the Ni/ZrO_2 (P) and Ni/ZrO_2 (H) catalysts was performed in a batch reactor at 210 °C and 5.0 MPa of H_2 pressure (Figure 8). The hydrogenation products were SA and GBL, with no other deep hydrogenation products or by-products being detected. Figure 8a shows that the Ni/ZrO_2 (P) exhibited a slightly higher MA conversion at the initial reaction stage. With a reaction time of 20 min, the conversion of MA for the Ni/ZrO_2 (P) catalyst was 51%, while that for the Ni/ZrO_2 (H) was 40%. MA conversion reached 100% for the Ni/ZrO_2 (P) and Ni/ZrO_2 (H) catalysts within 60 min. The product distribution was significantly different for the two catalysts, in the prolonged hydrogenation process which took 480 min. As shown in Figure 8b, the initial hydrogenation product was SA for the two catalysts before MA conversion reached 100%. As the time of the stream increased, the selectivity of the SA decreased gradually. This was accompanied by a gradual increase in the GBL selectivity over the Ni/ZrO_2 (H) catalyst. The GBL selectivity reached up to as much as 43.5% after 480 min. However, the SA selectivity decreased slightly over Ni/ZrO_2 (P) catalyst, and the GBL selectivity was only 2.8% after 480 min over Ni/ZrO_2 (P) catalyst.

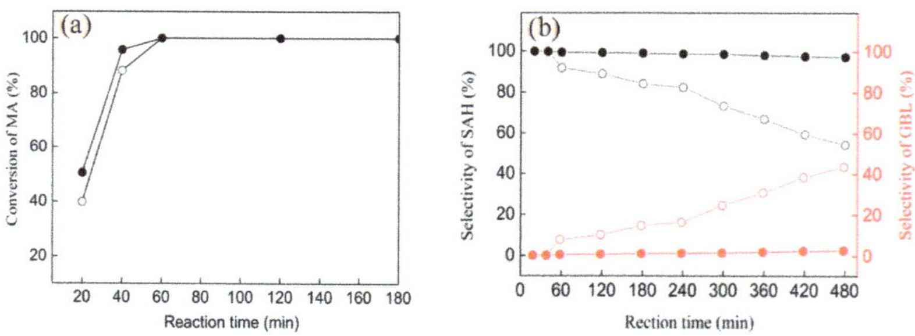

Figure 8. The MA conversion (**a**) and selectivity of SA and GBL (**b**) over the Ni/ZrO_2 (P) (solid symbols) and Ni/ZrO_2 (H) (open symbols) catalysts at 210 °C under 5 MPa for 480 min.

The above results showed that hydrogenation of MA to GBL was carried out in two successive reaction processes over the Ni/ZrO_2 catalysts—the hydrogenation of MA to SA, followed by the hydrogenation of C=O in SA to produce GBL. Before the MA conversion reached 100%, the hydrogen product was SA for the Ni/ZrO_2 (P) and Ni/ZrO_2 (H) catalysts. Furthermore, the Ni/ZrO_2 (P) catalyst exhibited a slightly higher C=C bond hydrogen activity. However, the C=O hydrogenation activity of the Ni/ZrO_2 (P) was much lower than that of the Ni/ZrO_2 (H) catalyst. A 43.5% yield of GBL was obtained over the Ni/ZrO_2 (H) catalyst at 210 °C and 5 MPa of H_2 pressure, while only 2.8% yield of GBL was obtained over the Ni/ZrO_2 (P) catalyst under the same reaction conditions. Even enhancing the reaction temperature or reaction pressure, the Ni/ZrO_2 (H) catalyst exhibited enhanced activity in C=O hydrogenation, whereas the Ni/ZrO_2 (P) catalyst still showed extremely low C=O hydrogenation activity (Table 4).

Table 4. MA conversion and GBL selectivity over the Ni/ZrO_2 (P) and Ni/ZrO_2 (H) catalysts under different temperature and pressure conditions for 480 min.

Catalysts	Temperature (°C)	Pressure (MPa)	Conv. (%)	GBL Selec. (%)
	210	5	100	2.8
Ni/ZrO_2 (P)	240	5	100	4.9
	210	7	100	3.2
	210	5	100	43.5
Ni/ZrO_2 (H)	240	5	100	60.6
	210	7	100	46.2

3.7. *H_2 Temperature-Programmed Desorption (H_2-TPD)*

For the hydrogenation reaction catalyzed by metallic nickel, the surface area of nickel plays a crucial role, as hydrogen dissociation always occurs on active metallic Ni^0 sites. To explore the reasons for the significant differences in C=O hydrogenation activity between the Ni/ZrO_2 (P) and Ni/ZrO_2 (H) catalysts, H_2-TPD characterization was conducted to investigate the activating hydrogen ability of the two catalysts. As shown in Figure 9, only one peak centered at 80 °C was detected for the Ni/ZrO_2 (H) catalyst. This was assigned to desorption of H which was weakly adsorbed on the Ni surface [51]. Compared to the Ni/ZrO_2 (H) catalyst, the Ni/ZrO_2 (P) exhibited two H desorption peaks: at lower temperature (~80 °C) and higher temperature (~206 °C) H desorption peaks. The new peak at the higher temperature originated from the more strongly chemisorbed H [51]. According to the literature [50], H_2-TPD peaks at temperatures below 300 °C can be attributed to the desorption of H from the surface of Ni. Our calculations showed that the amount of H desorption for Ni/ZrO_2 (P) was 1.12 times that of Ni/ZrO_2 (H) below 210 °C, which is consistent with the assumption that smaller Ni particles provide more hydrogen activation sites.

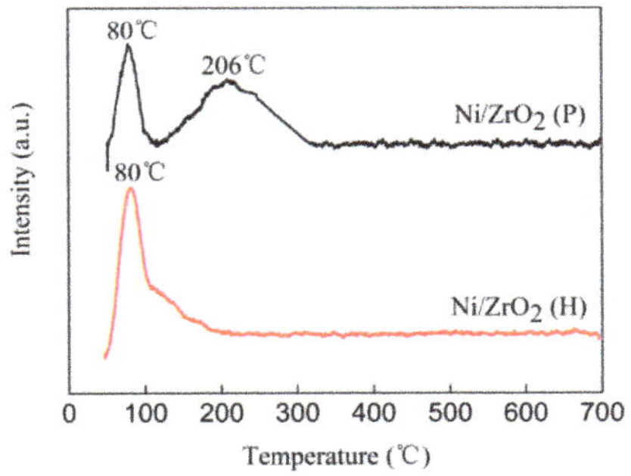

Figure 9. H_2-TPD profiles of Ni/ZrO_2 catalysts.

Generally, the more adsorption and activation hydrogen sites on the catalyst, the higher of the hydrogenation activity. The H_2-TPD characterization results showed that the Ni/ZrO_2 (P) catalyst possessed more hydrogen adsorption and activation sites, yet it exhibited only slight C=O hydrogenation activity. Therefore, it is surmised that the major reason for the different C=O hydrogenation activities between the two catalysts was their different adsorption and activation abilities for C=O.

3.8. *In-Situ FT-IR Spectra*

In order to explore the adsorption and activation abilities towards the C=O of the catalysts, in-situ FT-IR was investigated over two Ni/ZrO_2 catalysts using cyclohexanone as a probe molecule. In Figure 10, the peak centered at 1712 cm^{-1} was assigned to the C=O stretching vibration of pure cyclohexanone. Compared with the pure cyclohexanone, a significant redshift of the C=O stretching vibration peak occurred over the Ni/ZrO_2 (H) catalyst even down to 1627 cm^{-1}, while the peak corresponding to the C=O bond for the Ni/ZrO_2 (P) catalyst was weak and located at 1700 cm^{-1}. The two catalysts exhibited significantly different adsorption and activation properties towards C=O bonds. The much larger shift of the C=O bond indicated the weakening of the C=O bond and thereby the activation of them on surface of Ni/ZrO_2 (H) catalysts [52]. No obvious shift occurred on the Ni/ZrO_2 (P) catalyst, and the peak area was extremely small, indicating that the Ni/ZrO_2 (P) catalyst had very weak adsorption and activation abilities towards C=O.

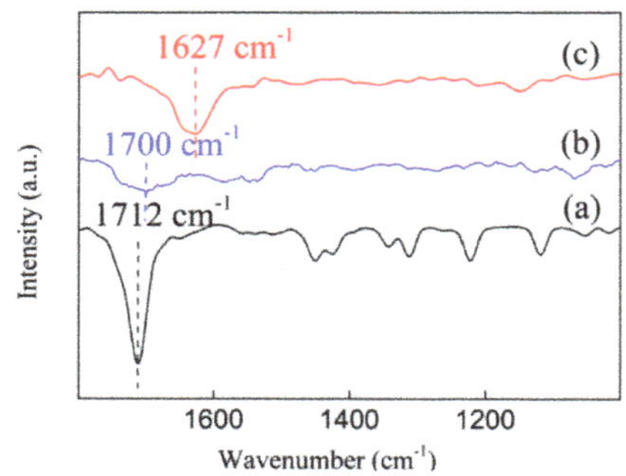

Figure 10. In-situ FT-IR of cyclohexanone (**a**) pure cyclohexanone, (**b**) cyclohexanone adsorbed on Ni/ZrO_2 (P), (**c**) cyclohexanone adsorbed on Ni/ZrO_2 (H).

4. Discussion

Generally, the catalytic performance of supported catalysts is intrinsically linked to the active metal sites and supports, including the active metal and metal-support interactions. As concerns hydrogenation reaction, it is well documented that H_2 can be dissociated over metal surfaces to generate active hydrogen, and a subsequent hydrogenation occurs with reactant molecules [53]. From this perspective, a catalyst with smaller metal nanoparticles should possess more accessible catalytically active sites and, consequently, exhibit higher hydrogenation activity. In this study, the Ni/ZrO_2 (P) catalyst exhibited stronger metal-support interaction than the Ni/ZrO_2 (H) catalyst because of its greater number of oxygen vacancies and low-coordinated oxygen ions on the surface. Thus, smaller Ni particles were obtained on Ni/ZrO_2 (P) catalyst. Furthermore, the H_2-TPD result illustrates that the Ni/ZrO_2 (P) catalyst possessed a higher metal Ni surface area and more hydrogen activation sites. In the MA hydrogenation reaction, the Ni/ZrO_2 (P) catalyst also exhibited a higher C=C hydrogenation activity, which was predictable and understandable. However, the C=O hydrogenation activity of the Ni/ZrO_2 (P) catalyst was much lower than that of the Ni/ZrO_2 (H) catalyst. A 43.5% yield of GBL was obtained over the Ni/ZrO_2 (H) catalyst, while 2.8% yield only (of GBL) was obtained over the Ni/ZrO_2 (P) catalyst under the same reaction conditions. Similar behavior was observed when we investigated the catalyst performances of MA hydrogenation over the Ni/ZrO_2(P) and Ni/ZrO_2(H) catalysts with 5 wt % nickel loading. The Ni crystalline size for Ni/ZrO_2 (P) catalyst with 5 wt % nickel loading was 9 nm while that for Ni/ZrO_2 (H) catalyst was 10 nm (Figure S1). Two catalyst possessed the similar crystalline size, whereas their catalytic performance was quite different (Table S1). The yield of GBL for Ni/ZrO_2 (P) catalyst with 5 wt % nickel loading was still very low, only 2.1%. While that for Ni/ZrO_2 (H) catalyst with 5 wt % nickel loading was 20.8%. The above results demonstrate that the C=O hydrogenation activities of the catalysts did not correlate well with the hydrogen activating ability of the catalyst or the Ni crystalline size. This strongly suggests that, in addition to the catalytic ability of metal Ni, other factors influenced the C=O hydrogenation activity for the studied ZrO_2-supported nickel catalyst system.

The results of in-situ FT-IR of adsorbed cyclohexanone showed that the Ni/ZrO_2 (H) catalyst was able to adsorb and activate C=O groups effectively, whereas the Ni/ZrO_2 (P) catalyst exhibited extremely weak adsorption and activation abilities for the same groups. The results of this study's experiments suggest that the superior catalytic activity of the Ni/ZrO_2 (H) catalyst in C=O hydrogenation can be attributed to its effective activation of the C=O group in the SA molecule. Hu et al. found that a large number of oxygen vacancies on Mn-containing spinel-supported copper catalyst contributed to the C=O hydrogenation [54]. Han et al. and Manyar et al. also observed similar results [55,56]. In our study, the Raman and XPS characterization results showed that the electronic properties of oxygen vacancies were significantly different on the surface of the Ni/ZrO_2 (P) and the Ni/ZrO_2 (H) catalyst. Surface oxygen vacancies on the Ni/ZrO_2 (P) catalyst exhibited relatively electron-rich properties while those on the Ni/ZrO_2 (H) catalyst showed relatively electron-deficient properties. Compared to relatively electron-rich oxygen vacancies, relatively electron-deficient oxygen vacancies were more likely to interact with lone pair electrons on the oxygen atoms of C=O groups, and so C=O groups could be activated. Given this, it can be deduced that relatively electron-deficient surface oxygen vacancies play a key role in promoting the hydrogenation of C=O groups. As concerns the Ni/ZrO_2 (H) catalyst, the oxygen vacancies promoted by Ni^0 located at the Ni/ZrO_2 interface exhibited lower charge densities and were more likely to adsorb and activate C=O groups in SA, thereby weakening the C=O bonds and lowering the energy requirement for hydrogenation and promoting C=O hydrogenation through their synergy with neighboring Ni particles.

Based on the above results, a plausible, simplified mechanism for C=O hydrogenation over Ni/ZrO_2 is proposed (Figure 11). Relatively electron-deficient oxygen vacancies on surface of Ni/ZrO_2 (H) catalyst activate the C=O bonds by accepting a lone pair of electrons from the oxygen atom of the C=O bonds, and thereby weakening the C=O bonds. Ni particles distributed near the relatively electron-deficient oxygen vacancies on the Ni/ZrO_2 (H) catalyst dissociated H_2 to produce active hydrogen and finish the C=O hydrogenation with the synergism of oxygen vacancies. As long as the relatively electron-rich oxygen vacancies on surface of Ni/ZrO_2 (P) catalyst cannot effectively activate the C=O bond, it will be difficult to achieve the C=O hydrogenation.

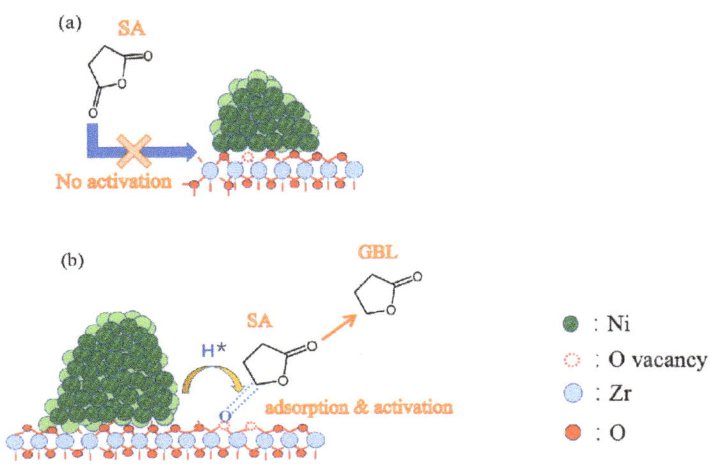

Figure 11. (**a**) Structure of nickel particles on the ZrO_2 (P). (**b**) Structure of nickel particles on the ZrO_2 (H) and hydrogenation of SA on the Ni/ZrO_2 (H).

5. Conclusions

This work showed an effective strategy for manipulating product selectivity in MA hydrogenation through regulating surface structures of Ni/ZrO_2 catalysts. The ZrO_2 (P) support, with more oxygen vacancies and low-coordinated oxygen ions on its surface, exhibited much stronger interactions with nickel species, which resulted in a small number of nickel species entering into oxygen vacancies and, thus, forming a Ni-O-Zr structure. This led to a decrease in the oxygen vacancy concentration and an increase of the average charge densities of the oxygen vacancies, which then produced high selectivity towards SA from MA hydrogenation. However, ZrO_2 (H), which had stable oxygen ions and fewer oxygen vacancies, showed weaker interactions with nickel species, resulting in large Ni particles being poorly dispersed on the Ni/ZrO_2 (H) and relatively electron-deficient oxygen vacancy generation promoted by Ni^0 particles. Thus, the Ni/ZrO_2 (H) exhibited a high selectivity towards GBL. The high C=O hydrogenation activities for the Ni/ZrO_2 (H) catalyst were attributed to the surface synergy between active metallic nickel species and relatively electron-deficient oxygen vacancies. These conclusions offer a new strategy for the design of high-efficiency selective hydrogenation catalysts applied to α, β-unsaturated aldehyde and ketone hydrogenation reactions by modulating the surface structure of ZrO_2 supports.

Supplementary Materials: The Supplementary Materials are available online at http://www.mdpi.com/2079-4991/9/3/406/s1. Figure S1: XRD patterns of Ni/ZrO_2 (P) and Ni/ZrO_2 (H) catalysts with 5 wt % nickel loading, Table S1: The catalytic performance in MA hydrogenation of Ni/ZrO_2 catalysts with 5 wt % nickel loading.

Author Contributions: Conceptualization, T.X. and Y.Z. (Yongxiang Zhao); methodology, L.Z.; software, M.Z. and W.Y.; validation, L.Z., T.W.; formal analysis, L.Z.; data curation, L.Z.; writing—original draft preparation, L.Z.; writing—review and editing, Y.Z. (Yin Zhang) and J.Z.; visualization, T.W. and H.L.; supervision, Y.Z. (Yongxiang Zhao) and Y.W.; project administration, Y.Z. (Yongxiang Zhao); funding acquisition, Y.Z. (Yongxiang Zhao)

Funding: This research was funded by the National Natural Science Foundation of China (21303097, 2167031118) and Key Projects of Shanxi Coal-based Low Carbon Joint Fund (U1710221).

Conflicts of Interest: The authors declare no conflicts of interest. The funders had no role in the design of the study; in the collection, analyses, or interpretation of data; in the writing of the manuscript; or in the decision to publish the results.

References

1. Trivedi, B.C.; Culbertson, B.M. *Maleic Anhydride*, 1st ed.; Plenum Press: New York, NY, USA, 2013; pp. 132–133.
2. Li, J.; Tian, W.P.; Shi, L. Hydrogenation of Maleic Anhydride to Succinic Anhydride over Ni/HY-Al_2O_3. *Ind. Eng. Chem. Res* **2010**, *49*, 11837–11840. [CrossRef]
3. Delbecq, F.; Sautet, P. A Density Functional Study of Adsorption Structures of Unsaturated Aldehydes on Pt (111): A Key Factor for Hydrogenation Selectivity. *J. Catal.* **2002**, *211*, 398–406. [CrossRef]
4. Lopez, A.; Bitzer, T.; Heller, T.; Richardson, N.V. Adsorption of maleic anhydride on Si (100)-2×1. *Surf. Sci.* **2001**, *477*, 219–226. [CrossRef]
5. Jung, S.M.; Godard, E.; Jung, S.Y.; Park, K.C.; Choi, J.U. Liquid-phase hydrogenation of maleic anhydride over Pd/SiO_2: Effect of tin on catalytic activity and deactivation. *J. Mol. Catal. A Chem.* **2003**, *198*, 297–302. [CrossRef]
6. Pallassana, V.; Neurock, M. First-Principles Periodic Density Functional Study of the Hydrogenation of Maleic Anhydride to Succinic Anhydride over Palladium (111). *J. Phys. Chem. B* **2000**, *104*, 9449–9459. [CrossRef]
7. Ponec, V. On the role of promoters in hydrogenations on metals; α-β unsaturated aldehydes and ketones. *Appl. Catal. A Gen.* **1997**, *149*, 27–48. [CrossRef]
8. Yu, Y.; Zhan, W.C.; Guo, Y.; Lu, G.Z.; Adjimi, S.; Guo, Y.L. Gas-phase hydrogenation of maleic anhydride to γ-butyrolactone over Cu-CeO_2-Al_2O_3 catalyst at atmospheric pressure: Effects of the residual sodium and water in the catalyst precursor. *J. Mol. Catal. A Chem.* **2014**, *395*, 392–397. [CrossRef]
9. Meyer, C.I.; Marchi, A.J.; Monzon, A.; Garetto, T.F. Deactivation and regeneration of Cu/SiO_2 catalyst in the hydrogenation of maleic anhydride. Kinetic modelling. *Appl. Catal. A Gen.* **2009**, *367*, 122–129. [CrossRef]
10. Feng, Y.H.; Yin, H.B.; Wang, A.L.; Xie, T.; Jiang, T.S. Selective hydrogenation of maleic anhydride to succinic anhydride catalyzed by metallic nickel catalysts. *Appl. Catal. A Gen.* **2012**, *425-426*, 205–212. [CrossRef]
11. Li, J.; Ren, Y.H.; Yue, B.; He, H.Y. Ni/Al_2O_3 catalysts derived from spinel $NiAl_2O_4$ for low-temperature hydrogenation of maleic anhydride to succinic anhydride. *Chin. J. Catal.* **2017**, *38*, 1166–1173. [CrossRef]

12. Li, J.; Tian, W.P.; Shi, L. Highly Active and Selective Nickel-Platinum Catalyst for the Low Temperature Hydrogenation of Maleic Anhydride to Succinic Anhydride and Synthesis of Succinic Acid at 40 °C. *Catal. Lett.* **2011**, *141*, 565–571. [CrossRef]
13. Guo, S.F.; Shi, L. Synthesis of succinic anhydride from maleic anhydride on Ni/diatomite catalysts. *Catal. Today* **2013**, *212*, 137–141. [CrossRef]
14. Meyer, C.I.; Regenhardt, S.A.; Bertone, M.E.; Marchi, A.J.; Garetto, T.F. Gas-Phase Maleic Anhydride Hydrogenation Over Ni/SiO_2-Al_2O_3 Catalysts: Effect of Metal Loading. *Catal. Lett.* **2003**, *143*, 1067–1073. [CrossRef]
15. Regenhardt, S.A.; Meyer, C.I.; Garetto, T.F.; Marchi, A.J. Selective gas phase hydrogenation of maleic anhydride over Ni-supported catalysts: Effect of support on the catalytic performance. *Appl. Catal. A Gen.* **2012**, *449*, 81–87. [CrossRef]
16. Cai, J.X.; Zhu, J.X.; Zuo, L.; Fu, Y.C.; Shen, J.Y. Effect of surface acidity/basicity on the selective hydrogenation of maleic anhydride to succinic anhydride over supported nickel catalysts. *Catal. Commun.* **2018**, *110*, 93–96. [CrossRef]
17. Huo, W.T.; Zhang, C.L.; Yuan, H.J.; Jia, M.J.; Ning, C.L.; Tang, Y.; Zhang, Y.; Luo, J.H.; Wang, Z.L.; Zhang, W.X. Vapor-phase selective hydrogenation of maleic anhydride to succinic anhydride over Ni/TiO_2 catalysts. *J. Ind. Eng. Chem.* **2014**, *20*, 4140–4145. [CrossRef]
18. Torresa, C.C.; Alderetea, J.B.; Mellab, C.; Pawelec, B. Maleic anhydride hydrogenation to succinic anhydride over mesoporous Ni/TiO_2 catalysts: Effects of Ni loading and temperature. *J. Mol. Catal. A Chem.* **2016**, *423*, 441–448. [CrossRef]
19. Hu, T.J.; Yin, H.B.; Zhang, R.C.; Wu, H.X.; Jiang, T.S.; Wada, Y.J. Gas phase hydrogenation of maleic anhydride to γ-butyrolactone by Cu-Zn-Ti catalysts. *Catal. Commun.* **2007**, *8*, 193–199. [CrossRef]
20. Yu, Y.; Guo, Y.L.; Zhan, W.C.; Guo, Y.; Wang, Y.Q.; Wang, Y.S.; Zhang, Z.G.; Lu, G.Z. Gas-phase hydrogenation of maleic anhydride to γ-butyrolactone at atmospheric pressure over Cu-CeO_2-Al_2O_3 catalyst. *J. Mol. Catal. A Chem.* **2011**, *337*, 77–81. [CrossRef]
21. Liao, X.; Zhang, Y.; Hill, M.; Xia, X.; Zhao, Y.X.; Jiang, Z. Highly efficient Ni/CeO_2 catalyst for the liquid phase hydrogenation of maleic anhydride. *Appl. Catal. A Gen.* **2014**, *488*, 256–264. [CrossRef]
22. Lin, Y.P.; Zhu, Y.F.; Pan, X.L.; Bao, X.H. Modulating the methanation activity of Ni by the crystal phase of TiO_2. *Catal. Sci. Technol.* **2017**, *7*, 2813–2818. [CrossRef]
23. Resende, K.A.; Braga, A.H.; Noronha, F.B.; Hori, C.E. Hydrodeoxygenation of phenol over Ni/$Ce_{1-x}Nb_xO_2$ catalysts. *Appl. Catal. B Environ.* **2019**, *245*, 100–113. [CrossRef]
24. Wang, Y.; Arandiyan, H.; Scott, J.; Dai, H.X.; Amal, R. Hierarchically Porous Network-Like Ni/Co_3O_4: Noble Metal-Free Catalysts for Carbon Dioxide Methanation. *Adv. Sustain. Syst.* **2018**, *2*, 1700119. [CrossRef]
25. Arandiyan, H.; Chang, H.Z.; Liu, C.X.; Peng, Y.; Li, J.H. Dextrose-aided hydrothermal preparation with large surface area on 1D single-crystalline perovskite $La_{0.5}Sr_{0.5}CoO_3$ nanowires without template: Highly catalytic activity for methane combustion. *J. Mol. Catal. A Chem.* **2013**, *378*, 299–306. [CrossRef]
26. Arandiyan, H.; Scott, J.; Wang, Y.; Dai, H.X.; Sun, H.Y.; Amal, R. Meso-Molding Three-Dimensional Macroporous Perovskites: A New Approach to Generate High-Performance Nanohybrid Catalysts. *ACS Appl. Mater. Interfaces* **2016**, *8*, 2457–2463. [CrossRef] [PubMed]
27. Zhu, Y.F.; Kong, X.; Cao, D.B.; Cui, J.L.; Zhu, Y.L.; Li, Y.W. The Rise of Calcination Temperature Enhances the Performance of Cu Catalysts: Contributions of Support. *ACS Catal.* **2014**, *4*, 3675–3681. [CrossRef]
28. De Souza, P.M.; Rabelo-Neto, R.C.; Borges, L.E.P.; Jacobs, G.; Davis, B.H.; Graham, U.M.; Resasco, D.E.; Noronha, F.B. Effect of Zirconia Morphology on Hydrodeoxygenation of Phenol over Pd/ZrO_2. *ACS Catal.* **2015**, *5*, 7385–7398. [CrossRef]
29. Samson, K.; Ś liwa, M.; Socha, R.P.; Gora-Marek, K.; Mucha, D.; Rutkowska-Zbik, D.; Paul, J-F.; Ruggiero-Mikołajczyk, M.; Grabowski, R.; Słoczynsk, J. Influence of ZrO_2 Structure and Copper Electronic State on Activity of Cu/ZrO_2 Catalysts in Methanol Synthesis from CO_2. *ACS Catal.* **2014**, *4*, 3730–3741. [CrossRef]
30. Rhodes, M.D.; Bell, A.T. The effects of zirconia morphology on methanol synthesis from CO and H_2 over Cu/ZrO_2 catalysts Part I. Steady-state studies. *J. Catal.* **2005**, *233*, 198–209. [CrossRef]
31. Zhang, X.P.; Zhang, Q.D.; Tsubaki, N.; Tan, Y.S.; Han, Y.Z. *Influence of Zirconia Phase on the Performance of Ni/ZrO_2 for Carbon Dioxide Reforming of Methane*; ACS Symposium Series; American Chemical Society: Washington, DC, USA, 2015; Volume 1194, pp. 135–153.

32. Jung, K.T.; Bell, A.T. The effects of synthesis and pretreatment conditions on the bulk structure and surface properties of zirconia. *J. Mol. Catal. A Chem.* **2000**, *163*, 27–42. [CrossRef]
33. Li, W.Z.; Huang, H.; Li, H.J.; Zhang, W.; Liu, H.C. Facile Synthesis of Pure Monoclinic and Tetragonal Zirconia Nanoparticles and Their Phase Effects on the Behavior of Supported Molybdena Catalysts for Methanol-Selective Oxidation. *Langmuir* **2008**, *24*, 8358–8366. [CrossRef] [PubMed]
34. Perry, C.H.; Lu, F.; Liut, D.W.; Alzyabt, B. Phonons and Phase Transitions in Zirconia. *J. Raman Spectrosc.* **1990**, *21*, 577–584. [CrossRef]
35. Bish, D.L.; Howard, S.A. Quantitative phase analysis using the Rietveld method. *J. Appl. Crystallogr.* **1988**, *21*, 86–91. [CrossRef]
36. Shukla, S.; Seal, S. Mechanisms of room temperature metastable tetragonal phase stabilisation in zirconia. *Int. Mater. Rev.* **2005**, *50*, 45–64. [CrossRef]
37. Yashima, M.; Ohtake, K.; Kakihana, M.; Arashis, H.; Yoshimura, M. Determination of tetragonal-cubic phase boundary of $Zr_{1-x}R_xO_{2-x/2}$ (R = Nd, Sm, Y, Er and Yb) by Raman scattering. *J. Phys. Chem. Solids* **1996**, *57*, 17–24. [CrossRef]
38. Li, C.; Li, M.J. UV Raman spectroscopic study on the phase transformation of ZrO_2, Y_2O_3-ZrO_2 and SO_4^{2-}/ZrO_2. *J. Raman Spectrosc.* **2002**, *33*, 301–308. [CrossRef]
39. Shi, L.; Tin, K.C.; Wong, N.B. Thermal stability of zirconia membranes. *J. Mater. Sci.* **1999**, *34*, 3367–3374. [CrossRef]
40. Lopez, E.F.; Escribano, V.S.; Panizza, M.; Carnascialic, M.M.; Busca, G. Vibrational and electronic spectroscopic properties of zirconia Powders. *J. Mater. Chem.* **2001**, *11*, 1891–1897. [CrossRef]
41. Thackeray, D.P.C. The Raman spectrum of zirconium dioxide. *Spectrochim. Acta Part A Mol. Biomol. Spectrosc.* **1974**, *30*, 549–550. [CrossRef]
42. Yashima, M.; Ohtake, K.; Arashi, H.; Kakihana, M.; Yoshimura, M. Determination of cubic-tetragonal phase boundary in $Zr_{1-X}Y_XO_{2-X/2}$ solid solutions by Raman spectroscopy. *J. Appl. Phys.* **1993**, *74*, 7603–7605. [CrossRef]
43. Chen, S.Q.; Li, L.P.; Hu, W.B.; Huang, X.S.; Li, Q.; Xu, Y.S.; Zuo, Y.Y.; Li, G.S. Anchoring High-Concentration Oxygen Vacancies at Interfaces of CeO_{2-x}/Cu toward Enhanced Activity for Preferential CO Oxidation. *ACS Appl. Mater. Interfaces* **2015**, *7*, 22999–23007. [CrossRef] [PubMed]
44. Puigdollers, A.R.; Illas, F.; Pacchioni, G. Structure and Properties of Zirconia Nanoparticles from Density Functional Theory Calculations. *J. Phys. Chem. C* **2016**, *120*, 4392–4402. [CrossRef]
45. Tosoni, S.; Chen, H.Y.T.; Pacchioni, G. A DFT study of Ni clusters deposition on titania and zirconia (101) surfaces. *Surf. Sci.* **2016**, *646*, 230–238. [CrossRef]
46. Yang, F.F.; Liu, D.; Zhao, Y.T.; Wang, H.; Han, J.Y.; Ge, Q.F.; Zhu, X.L. Size Dependence of Vapor Phase Hydrodeoxygenation of m-Cresol on Ni/SiO_2 Catalysts. *ACS Catal.* **2018**, *8*, 1672–1682. [CrossRef]
47. Rahman, M.A.; Rout, S.; Thomas, J.P.; McGillivray, D.; Leung, K.T. Defect-Rich Dopant-Free ZrO_2 Nanostructures with Superior Dilute Ferromagnetic Semiconductor Properties. *J. Am. Chem. Soc.* **2016**, *138*, 11896–11906. [CrossRef] [PubMed]
48. Chang, S.J.; Li, M.; Hua, Q.; Zhang, L.J.; Ma, Y.S.; Ye, B.J.; Huang, W.X. Shape-dependent interplay between oxygen vacancies and Ag-CeO_2 interaction in Ag/CeO_2 catalysts and their influence on the catalytic activity. *J. Catal.* **2012**, *293*, 195–204. [CrossRef]
49. Puigdollers, A.R.; Schlexer, P.; Tosoni, S.; Pacchioni, G. Increasing Oxide Reducibility: The Role of Metal/Oxide Interfaces in the Formation of Oxygen Vacancies. *ACS Catal.* **2017**, *7*, 6493–6513. [CrossRef]
50. Hengne, A.M.; Samal, A.K.; Enakonda, L.R.; Harb, M.; Gevers, L.E.; Anjum, D.H.; Hedhili, M.N.; Saih, Y.; Huang, K.W.; Marie Basset, J. Ni-Sn-Supported ZrO_2 Catalysts Modified by Indium for Selective CO_2 Hydrogenation to Methanol. *ACS Omega* **2018**, *3*, 3688–3701. [CrossRef]
51. Martensson, A.S.; Nyberg, C.; Andersson, S. Adsorption of hydrogen on a stepped nickel surface. *Surf. Sci* **1988**, *205*, 12–24. [CrossRef]
52. Cui, J.L.; Tan, J.J.; Zhu, Y.L.; Cheng, F.Q. Aqueous Hydrogenation of Levulinic Acid to 1,4-Pentanediol over Mo-Modified Ru/Activated Carbon Catalyst. *ChemSusChem* **2018**, *11*, 1–6. [CrossRef] [PubMed]
53. Bertero, N.M.; Trasarti, A.F.; Apesteguía, C.R.; Marchi, A.J. Solvent effect in the liquid-phase hydrogenation of acetophenone over Ni/SiO_2: A comprehensive study of the phenomenon. *Appl. Catal. A Gen.* **2011**, *394*, 228–238. [CrossRef]

54. Hu, Q.; Yang, L.; Fan, G.L.; Li, F. Hydrogenation of biomass-derived compounds containing a carbonyl group over a copper-based nanocatalyst: Insight into the origin and influence of surface oxygen vacancies. *J. Catal.* **2016**, *340*, 184–195. [CrossRef]
55. Han, J.S.; Kim, Y.H.; Jang, H.S.; Hwang, S.Y.; Jegal, J.; Kim, J.W.; Lee, Y.S. Heterogeneous zirconia-supported ruthenium catalyst for highly selective hydrogenation of 5- hydroxymethyl-2-furaldehyde to 2,5-bis(hydroxymethyl)furans in various n-alcohol solvents. *RSC Adv.* **2016**, *6*, 93394–93397. [CrossRef]
56. Manyar, H.G.; Paun, C.; Pilus, R.; Rooney, D.W.; Thompson, J.M.; Hardacre, C. Highly selective and efficient hydrogenation of carboxylic acids to alcohols using titania supported Pt catalysts. *Chem. Commun.* **2010**, *46*, 6279–6281. [CrossRef] [PubMed]

nanomaterials

MDPI

Communication

The Synergistic Effect of Acidic Properties and Channel Systems of Zeolites on the Synthesis of Polyoxymethylene Dimethyl Ethers from Dimethoxymethane and Trioxymethylene

Jianbing Wu [1,*], Sen Wang [2,*], Haitao Li [1], Yin Zhang [1], Ruiping Shi [2] and Yongxiang Zhao [1,*]

1 Engineering Research Center of Ministry of Education for Fine Chemicals, Shanxi University, Taiyuan 030006, China

2 State Key Laboratory of Coal Conversion, Institute of Coal Chemistry, Chinese Academy of Sciences, Taiyuan 030001, China

* Correspondence: wujianbing@sxu.edu.cn (J.W.); wangsen@sxicc.ac.cn (S.W.); yxzhao@sxu.edu.cn (Y.Z.)

Received: 16 July 2019; Accepted: 19 August 2019; Published: 23 August 2019

Abstract: A series of zeolites with different topology structures, including SAPO-34, SUZ-4, ZSM-5, USY, MOR, and beta, were used to synthesize polyoxymethylene dimethyl ethers ($PODE_n$) from dimethoxymethane (DMM) and trioxymethylene (TOM). The influence of acidic properties and channel systems were studied by activity evaluation, characterization, and theoretical calculation. The results confirmed that pore mouth diameter larger than a TOM molecule was an essential prerequisite for the synthesis of $PODE_n$ over zeolites, and the synergistic effect between medium-strong Brønsted acid sites (Brønsted MAS) and the maximal space of zeolites available determined the catalytic performance of all studied zeolites. DMM and TOM were firstly decomposed into methoxymethoxy groups (MMZ) and monomer CH_2O over Brønsted MAS. Subsequently, the steric constraint of the maximum included sphere, with an appropriate size in zeolite channels, can promote the combination of CH_2O and MMZ to form transition species $ZO(CH_2O)_nCH_3$, which reacted with the methyl-end group to form $PODE_n$ over Brønsted MAS. Moreover, the reaction temperature showed different effects on the product selectivity and distribution, which also mainly depends on the size of the maximum space available in zeolite channels.

Keywords: zeolites; polyoxymethylene dimethyl ethers; dimethoxymethane; trioxymethylene; Brønsted acid sites; the maximum included sphere; steric constraint

1. Introduction

The growing consumption of petroleum fuels is one of the major reasons for air pollution and environmental imbalance. Many efforts have been made to reduce exhaust gas and solid particulate matter from commercial vehicles. The identification of a highly efficient and clean fuel additive that could enhance diesel oil combustion efficiency and reduce pollutant emissions would be an economical and practical solution. Recently, polyoxymethylene dimethyl ethers ($PODE_n$, CH_3–O–(CH_2–O)$_n$–CH_3, $n > 1$) have exhibited much potential as diesel additives, due to their high oxygen content, cetane number, good miscibility, and absence of C–to–C bonds [1–5]. Compared with CH_3OH (MeOH), CH_3OCH_3 (DME), $CH_3OCH_2OCH_3$ (DMM), and $CH_3OCOOCH_3$ (DMC), $PODE_n$, especially $PODE_{2–8}$, can directly blend into traditional diesel without any fuel system modifications and can reduce the release of NO_x and particulate emissions effectively [6–8]. Accordingly, the synthesis of $PODE_n$ has gained much research interest.

$PODE_n$ molecules are primarily made up of a methyl-end group and oxymethylene group, which are usually obtained from methanol (MeOH) or dimethoxymethane (DMM) and formaldehyde (FA) or polyformaldehydes (PF_n), respectively [4]. A previous investigation of the synthesis of $PODE_n$ was carried out in homogeneous catalytic systems with H_2SO_4 or CF_3SO_3H as catalysts, which resulted in high corrosivity, separation difficulties, and potential environmental damage [9]. Consequently, various heterogeneous catalysts, including ion exchange resins [10,11], solid oxides [12–14], and zeolites [15–20], were evaluated for $PODE_n$ synthesis, due to their advantages in tunable acidity and separation properties.

Our previous work found that the MCM-22 zeolite can catalyze MeOH and trioxymethylene (TOM) to synthesize $PODE_n$, with a $PODE_{3-8}$ selectivity of 29% at 120 °C [17]. However, due to the water and additional byproducts that were produced with MeOH and TOM or PF as raw materials, which are detrimental to the desired product properties [21], $PODE_n$ synthesis with DMM and TOM was also studied in detail. A $PODE_{2-8}$ selectivity of 88.5% and a TOM conversion of 85.3% were obtained at 120 °C over the ZSM-5 catalyst with a Si/Al molar ratio of 580 [15]. In general, the synthesis of $PODE_n$ from DMM and TOM is most likely the result of the following reversible reactions (Figure 1): The TOM ring decomposes to monomer CH_2O, after which CH_2O is inserted into DMM one by one to generate $PODE_n$ [4]. Although different zeolites have been developed to synthesize $PODE_n$ [15,17,22], most research has mainly focused on acidity regulation, whereas the influence of zeolite frameworks and synergistic effect between acidic properties and channel systems remains to be understood.

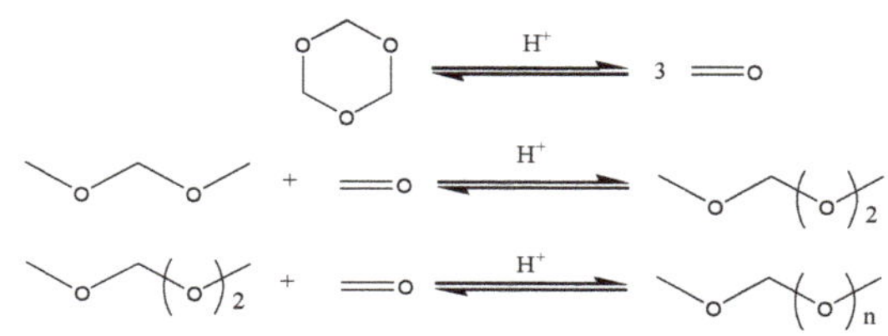

Figure 1. Reaction routes of polyoxymethylene dimethyl ethers ($PODE_n$) synthesis from dimethoxymethane (DMM) and trioxymethylene (TOM) [4].

The aim of the present investigation was to evaluate the effects of zeolite acidic characteristics and channel systems on the synthesis of $PODE_n$ from DMM and TOM, as well as to further evaluate the synergistic effect between these two factors. SAPO-34, SUZ-4, ZSM-5, MOR, USY, and beta zeolites were selected as catalysts, on the basis of similar Si/Al ratios and different framework types. Experiments were carried out to determine the effects of acidic characteristics and channel systems at different temperatures on the feedstock conversion, product selectivity, and product distribution. The observed effects are interpreted in the light of a series of activity evaluations, characterizations, and theoretical calculations.

2. Materials and Methods

Boehmite(P-D-03) was purchased from Aluminum Corporation of China. Carbon-white (Cab-o-sil M5) was bought from Cabot Co. Silica sol (40.5 wt% of SiO_2) was purchased from Qingdao Haiyang Chem. Co., Ltd. Triethylamine, phosphoric acid(85 wt %), sodium aluminate (41 wt% of Al_2O_3, 41 wt% of Na_2O), sodium hydroxide (96 wt%), potassium hydroxide, tetraethylammonium hydroxide (25 wt%), tetrapropylammonium hydroxide(25%) were all obtained from Sinopharm Chemical Reagent Co., Ltd. All the reagents were used as received without further purification.

SAPO-34, SUZ-4, ZSM-5, and MOR zeolites were synthesized following the procedures reported in the literature [23–26]. USY (Na-FAU, Si/Al = 12) and beta zeolites (Na-BEA, Si/Al = 10) were obtained from commercial manufacturers (XFNANO, NanJing, China). All the as-synthesized zeolites were calcined at 550 °C for 10 h under static air to remove the organic template, followed by two rounds of ion exchange with NH_4NO_3 solution (1 M) at 80 °C. Afterwards, the zeolites were washed with deionized water, dried at 110 °C, and then calcined under static air at 550 °C for 10 h to obtain the

H-form samples. Further experimental details regarding the characterizations and calculations are provided in the "Supplementary Materials".

The catalytic performance of zeolites was tested in a stainless steel autoclave lined with Teflon. In each run, DMM (3.8 g), TOM (2.25 g), and catalysts (0.3 g) were loaded into the 25 mL autoclave. The mixture was stirred continuously for 45 min and the reaction temperature was controlled between 30 °C and 150 °C. All products after the reaction consisted of $PODE_n$, unreacted reactants and byproducts were measured by gas chromatography (GC) and quantitatively analyzed by an internal standard method in which decane was the internal standard. In order to further ensure the accuracy of the results, products were also detected by gas chromatography–mass spectrometry (GC–MS). The conversions of DMM and TOM, denoted as x_{DMM} and x_{TOM}, were calculated by Equations (1) and (2).

$$x_{DMM} = (m_{\mathrm{DMM,feed}} - m_{\mathrm{DMM,product}})/m_{\mathrm{DMM,feed}} \times 100\% \quad (1)$$

$$x_{\mathrm{TOM}} = (m_{\mathrm{TOM,feed}} - m_{\mathrm{TOM,product}} - m_{\mathrm{FA,product}})/m_{\mathrm{TOM,feed}} \times 100\%. \quad (2)$$

The mass selectivity of products was determined by Equation (3)

$$M_i = \mathrm{m}_{\mathrm{i,product}}/\Sigma m_{\mathrm{i}} \times 100\%. \quad (3)$$

For example, $M_{PODE2-8} = m_{\mathrm{PODE2-8,product}}/\Sigma m_{\mathrm{i}} \times 100\%$, where $m_{\mathrm{PODE2-8}}$ is the mass of PODE_{2-8} and Σm_i is the mass of all the trapped liquid products after the reaction.

3. Results and Discussion

The X-ray diffraction (XRD) patterns (Figure S1) illustrate that the as-synthesized zeolites were well crystallized with the characteristic peaks of corresponding topology structure [23–26]. SEM images (Figure S2) further confirm it. The acidities of H-form zeolites were measured by temperature-programmed ammonia desorption (NH_3-TPD) and infrared spectroscopy of pyridine adsorption (Py-IR) (Figure 2). The Si/Al ratios, surface area, pore volume, detailed weak acid sites (WAS), medium-strong acid sites (MAS), Brønsted acid sites (BAS), and Lewis acid sites (LAS) were calculated (Table 1). Two ammonia desorption peaks were observed in the NH_3-TPD profiles (Figure 2a), one at 120–250 °C and the other at 250–550 °C, corresponding to WAS and MAS, respectively. The desorption temperature of WAS were generally similar for all the samples; however, there was a notable difference in the desorption temperature of MAS: The higher the desorption temperature, the stronger was the acidic strength. The acidic strength of MAS gradually decreased in the following order: MOR > SUZ-4 > ZSM-5 > SAPO-34 > USY > beta. Furthermore, the peak intensity represents the density of acid sites (Table 1).

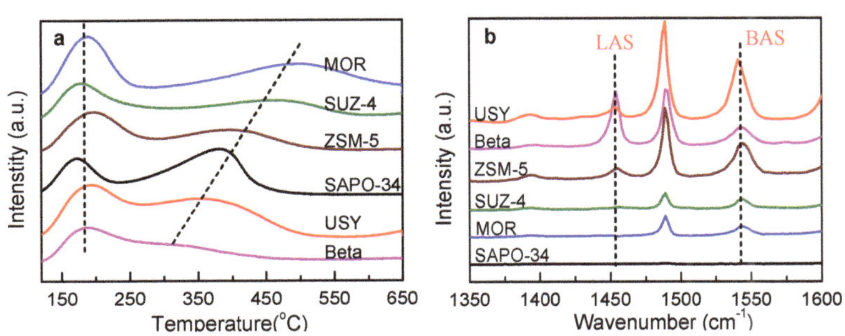

Figure 2. The temperature-programmed ammonia desorption (NH_3-TPD) profiles (**a**) and pyridine adsorption (Py-IR profiles) (**b**) of various zeolites.

The Py-IR of all the zeolites at different temperatures in the region of 1350–1600 cm^{-1} (Figure S3) with the profiles at 250 °C (Figure 2b) were assessed. Two main characteristic bands of pyridine adsorbed on BAS (1540 cm^{-1}) and LAS (1450 cm^{-1}) were observed [27–29]. The amounts of BAS and LAS (Table 1) and their distribution (Figure 2a,b) differ distinctly from the WAS and MAS of the NH_3-TPD characterization, especially for SUZ-4, MOR, and SAPO-34. The total acid site value of SUZ-4 was 0.840 mmol g^{-1}, but only 0.073 mmol g^{-1} for BAS and 0.013 mmol g^{-1} for LAS. The total acid site value of SAPO-34 was 1.07 mmol g^{-1}, whereas no BAS or LAS were detected. For MOR, the total acid site value was 1.63 mmol g^{-1} and BAS was only 0.12 mmol g^{-1}. This difference is mainly due to variations in zeolite topology structures. SUZ-4 has a channel system that includes 10-membered ring (MR) channels parallel to the unit cell c-axis, which are intersected by two arrays of 8-MR channels running in the plane. SAPO-34 contains large cavities, but these cavities are connected by small 8-MR pore mouths. MOR contains two essentially different channels: Large 12-MR channels and small 8-MR channels that run in parallel and are connected by 8-MR openings. The diameter of 8-MR is much smaller than that of the large pyridine molecule (5.3 Å), and thus, prevents its diffusion into zeolite channels [30]. Therefore, parts of BAS and LAS that are located in 8-MR channels cannot be determined by pyridine adsorption.

Table 1. Physicochemical and acidic properties of various zeolite samples.

Samples	Physicochemical Properties			Acidity by Strength [d] (mmol g^{-1})			Acidity by Type [e] (mmol g^{-1})	
	Si/Al Ratio [a]	S_{BET} (m^2 g^{-1}) [c]	V_{pore} (cm^3 g^{-1})	Weak	Medium—Strong	Total	Brønsted	Lewis
USY	10	641	0.372	0.40	0.24	0.64	0.50	0.092
Beta	12	500	0.346	0.53	0.23	0.76	0.25	0.40
ZSM-5	17	368	0.231	0.52	0.47	0.99	0.47	0.086
SUZ-4	6	348	0.337	0.40	0.44	0.84	0.073	0.013
SAPO-34	0.085 [b]	531	0.302	0.27	0.80	1.07	–	–
MOR	15	500	0.259	0.63	1.00	1.63	0.12	0

Notes: [a]: The Si/Al atomic ratio was determined by inductively coupled plasma atomic emission spectroscopy (ICP-AES); [b]: This value was obtained by determining the Si/(Al + P) atomic ratio; [c]: Surface area (S_{BET}) and total pore volume (V_{pore}) were obtained from nitrogen sorption data; [d]: density of the acid sites, assorted according to acidic strength, determined by NH_3-TPD; medium-strong: NH_3 desorbed at 250–550 °C; weak: NH_3 desorbed at 120–250 °C; [e]: Density of the acid sites, assorted according to the acidic type, determined by Py-IR.

A series of preliminary tests with ZSM-5 as a representative catalyst (Figures S4–S7) illustrated that a feed DMM/TOM molar ratio of 2 and a catalyst amount of 5 wt% for 45 min are appropriate for the synthesis of $PODE_n$ [15] and that the zeolite particle size has little effect on catalytic performance. Based on this, the catalytic performance of different zeolites was evaluated as functions of reaction temperature (Figure 3). Beta, ZSM-5, and MOR displayed very similar trends in selectivity to $PODE_{2-8}$ (Figure 3c), reaching a maximum selectivity of 83%–87% in the range of 70–90 °C. Meanwhile, TOM conversion increased quickly before 90 °C and then continued to rise slowly with reaction temperature (Figure 3a). DMM conversion increased initially, reaching a maximum value at 90 °C, and then gradually declined (Figure 3b). The byproduct selectivity of the above three zeolites (Figure 3e–h) were also very similar over the operating temperature range. MeOH and HCHO (FA) decreased, whereas $HCOOCH_3$ (MF) and CH_3OCH_3 (DME) increased and turned into primary products with increasing reaction temperatures.

In contrast, the catalytic performances of SAPO-34, SUZ-4, and USY exhibited great variability. For SAPO-34 and SUZ-4, a reaction occurred only when the temperature was higher than 100 °C. SAPO-34 exhibited poor catalytic activity, only 47.5% $PODE_{2-8}$ selectivity and 13.8% $PODE_{2-8}$ yield were obtained at 150 °C. SUZ-4 was slightly superior to SAPO-34, with 68.8% $PODE_{2-8}$ selectivity at 110 °C and 48.8% yield at 130 °C. Unlike SAPO-34 and SUZ-4, the catalytic performance of USY gradually increased with a rise of reaction temperature from 30 °C to 150 °C, and ultimately 80.0% $PODE_{2-8}$ selectivity and 72.4% yield were obtained at 150 °C. This result implies that USY can achieve a similar catalytic performance as beta, ZSM-5, and MOR, but only at higher temperatures.

Furthermore, a detailed PODE$_n$ (n = 2–11) distribution over USY and ZSM-5 was assessed (Figure 4). As the high Si/Al ratio of HZSM-5 proved effective in the synthesis of PODE$_n$ from DMM and TOM [15], ZSM-5 with a Si/Al ratio of 267 was selected as a research object for comparison. The bar diagrams from bottom to top represent the changing trend of PODE$_2$ to PODE$_{11}$ selectivity (Figure 4). The selectivity of PODE$_n$ gradually decreased with the increase of n from 2 to 11 at each reaction temperature. Increasing the Si/Al ratio from 17 to 267 not only increased PODE$_n$ selectivity between 70 °C and 90 °C, but also delayed the rapid selectivity decline at higher temperatures. This indicates that appropriately decreasing the Al content of ZSM-5 could facilitate increasing long-chain PODE$_n$ selectivity and suppressing certain side reactions at high temperatures. Furthermore, USY exhibits poor catalytic performance at low temperatures, but exhibits a similar PODE$_n$ selectivity between 110 °C and 150 °C as ZSM-5 with a high Si/Al ratio.

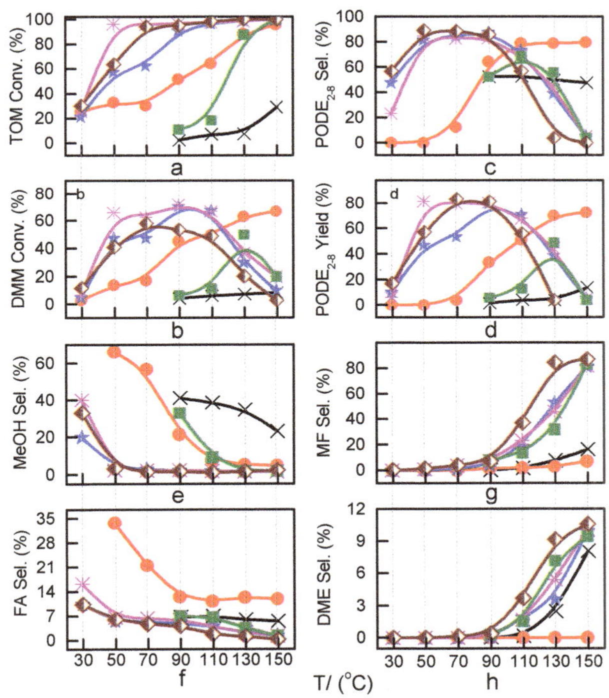

Figure 3. Reactant conversion and product selectivity over different zeolites. ●: USY; ✳: Beta; ◆: ZSM-5; ■: SUZ-4; ×: SAPO-34; ★: MOR; (**a**): TOM conversion, (**b**): PODE$_{2\text{-}8}$ selectivity, (**c**): DMM conversion, (**d**): PODE$_{2\text{-}8}$ yield, (**e**): MeOH selectivity; (**f**): FA selectivity; (**g**): MF selectivity; (**h**): DME selectivity.

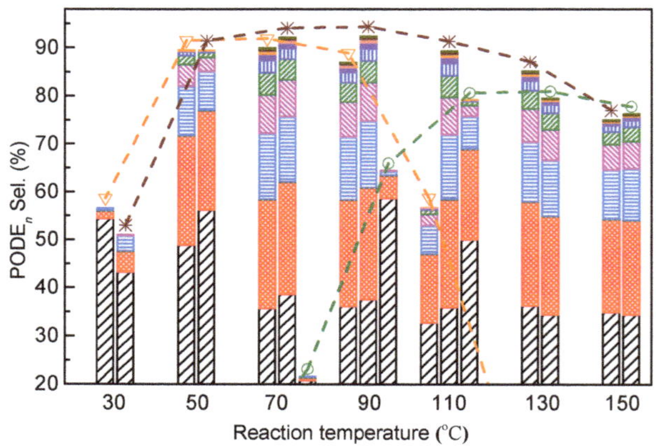

Figure 4. The PODE$_n$ product distribution over ZSM-5 and USY. ▽: ZSM-5 (Si/Al = 17), ✱: ZSM-5 (Si/Al = 267), ○: USY (Si/Al = 10), black bar: PODE$_2$, red bar: PODE$_3$, sky-blue bar: PODE$_4$, magenta bar: PODE$_5$, olive bar: PODE$_6$, blue bar: PODE$_7$, violet bar: PODE$_8$, orange bar: PODE$_9$, wine bar: PODE$_{10}$, dark yellow bar: PODE$_{11}$.

The synthesis of PODE$_n$ is an acid-catalyzed reaction. Acidic properties, including the density, strength, circumstance, and type of acid sites, directly affect the catalytic performance of zeolites. All the zeolites in this study were relatively similar regarding the desorption temperature of WAS, but exhibited tremendous variations in MAS (Figure 3a). This suggests that catalytic performance strongly depends on MAS. Our previous work has also demonstrated that sufficient Brønsted MAS can effectively promote TOM dissociation and PODE$_n$ chain propagation [15].

Although SAPO-34 contains close to 0.80 mmol g^{-1} MAS, no Brønsted acids were detected that could cause the very low PODE$_{2-8}$ yield. Similarly, SUZ-4 contains only 0.073 mmol g^{-1} Brønsted acids that result in its poor catalytic activity. In comparison, beta, ZSM-5, and MOR exhibit better catalytic performance than SAPO-34 and SUZ-4, which should be attributed to more sufficient BAS. However, it is surprising that USY exhibits completely different catalytic performance that is unlike any other zeolite. The acidic property of USY is very close to beta in regards to acid strength and acid density. The desorption temperature of MAS was approximately 350 °C, and acid density of USY and Beta were 0.24 mmol g^{-1} and 0.23 mmol g^{-1}, respectively. Moreover, the Brønsted acid density of USY was more than beta. These results indicate that USY should exhibit similar or better catalytic activity than beta in the same conditions, but obvious differences exist between them. The large differences in catalytic performance of the above-mentioned zeolite catalysts suggest that Brønsted MAS are not the only influential factors in the synthesis of PODE$_n$. In other words, zeolite acidic properties cannot explain the observed differences in zeolite catalytic performance fully. The following paragraphs will discuss that these differences can be ascribed to differences in channel systems.

Detailed zeolite characteristics (Table 2) were obtained from the International Zeolite Association (IZA) database of zeolite channel systems and related literature [31,32]. Generally, USY, beta, and MOR have 12-ring pores and are usually considered large-pore zeolites; ZSM-5 and SUZ-4 have 10-ring pores and are considered medium-pore zeolites; and SAPO-34 has 8-ring pores and is considered a small-pore zeolite. The maximum included sphere diameters (Table 2) in each zeolite framework type are considered the largest hard sphere or the maximum space available and are usually located inside channel intersections or cage structures and are stationary [32,33].

Table 2. Structural properties of different zeolite framework types.

Samples	Channel System		
	Pore Diameter (nm)	**Channel Structure**	**Maximum Included Sphere Diameter (Å)**
USY	0.74 × 0.74	3D, 12-ring	11.18
Beta	0.66 × 0.67	3D, 12-ring	6.62
ZSM-5	0.53 × 0.56	3D, 10-ring	6.30
SUZ-4	0.41 × 0.52	3D, 10-ring	6.27
SAPO-34	0.38 × 0.38	3D, 8-ring	7.31
MOR	0.65 × 0.70	1D, 12-ring	6.64

The pore mouth size of different zeolites is believed to play an important role in the reaction that occurs in the zeolite pore. BAS were not detected in SAPO-34, which we have ascribed to the fact that the pyridine molecule is too large (5.3 Å) to pass through the pore mouth (3.8 Å). This also holds true for the synthesis of $PODE_n$. Theoretical calculations revealed two conformations of the TOM molecule (Table 3), whose sizes are both larger than 4 Å and thus also larger than the pore mouth size of SAPO-34 (3.8 Å). This indicates that the TOM molecule cannot enter into the SAPO-34 channel and thus subsequent TOM decomposition and $PODE_n$ synthesis cannot occur. That is to say, the small pore mouth size prevents catalytic performance of SAPO-34, while the poor activity of SAPO-34 at high temperatures may be attributed to DMM decomposition or disproportionation, resulting in the formation of $PODE_2$ and other byproducts [33]. This was also illustrated by the DMM disproportionation over H-ZSM-5 at different temperatures (Figure S8). These results suggest that the synthesis of $PODE_n$ occurs in the zeolite pore and that a pore diameter that is larger than a TOM molecule is an essential prerequisite for the reaction.

However, pore mouth size is not the decisive factor resulting in the difference in $PODE_n$ selectivity among USY, ZSM-5, MOR, and beta, whose pore mouth sizes are much larger than the TOM molecule. The maximal included sphere sizes in beta, ZSM-5, MOR, and USY are 6.62 Å, 6.30 Å, 6.64 Å, and 11.18 Å, respectively. The maximum space available is nearly uniform for these zeolites, except USY, which is at least 1.68 times larger than the other three zeolites [33]. This suggests that the maximum included sphere in the zeolite channel may be another key factor. Taking into consideration that the adsorption of reactants onto catalysts is the initial step necessary for the synthesis of $PODE_n$, models of ZSM-5 and USY were built. The adsorption energy of DMM and TOM and the electrostatic interactions of reacting molecules with the ZSM-5 and USY frameworks were investigated with the density functional theory method. The acid site is located in the intersection of ZSM-5 or the supercage of USY, which provides the maximal space available to accommodate various guest molecules.

Table 3. The theoretical molecular size of TOM.

TOM Chair Conformation		**TOM Boat Conformation**	
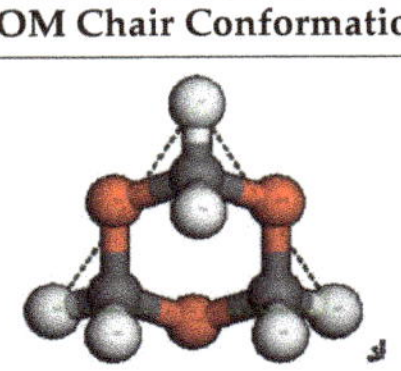		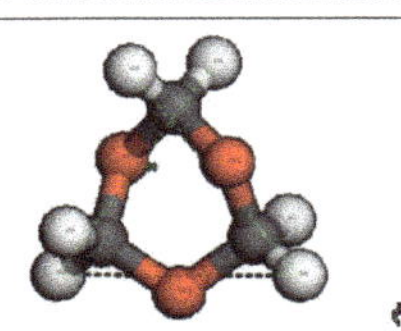	
H4-H6	4.033 Å	H4-H9	4.004 Å
H4-H7	4.033 Å	H8-H5	4.004 Å
H6-H7	4.033 Å	H8-H9	4.038 Å

The adsorption states of DMM, TOM, and their combination, as well as adsorption energies of ZSM-5 and USY, were modelled (Table 4). ZSM-5 exhibits much higher adsorption energies than USY, regardless of the presence of a single reactant or mixture, suggesting that the interactions between reactants and ZSM-5 frameworks are more intense and that the synthesis of $PODE_n$ should be sensitive to spatial constraints. The reaction of DMM with the Brønsted acidic protons of the zeolite can lead to the formation of methoxymethyl species (CH_3OCH_2OZ MM-Z), and TOM can decompose to CH_2O [15,33]. Celik et al. [33] proposed that the steric constraint of the pore walls in the channel intersections of ZSM-5 leads to high DMM disproportionation rates by promoting the methylene hydrogens of the DMM molecule to interact with the MM-Z. Conversely, the supercage structures of USY allow the hydrogen donor to remain far from the acceptor, so that the driving force for hydrogen transfer is not as strong. The steric constraint of the pore walls may also influence TOM decomposition and $PODE_n$ synthesis in a similar manner. The highest $PODE_{2-8}$ selectivity over ZSM-5 was obtained at 70–90 °C (Figure 3c); however, the main products over USY are MeOH and FA under the same conditions that come from DMM decomposition and TOM decomposition, respectively. This phenomenon may be explained by the fact that the steric constraint of the ZSM-5 channel intersections can preferentially promote the combination of MM-Z and CH_2O from DMM and TOM to form long-chain $ZO(CH_2O)_nCH_3$, which further combine with the methyl-end group over Brønsted MAS, resulting in the formation of $PODE_n$. However, the driving force of USY for the combination of intermediate species is not as strong due to its large size, resulting in poor $PODE_n$ selectivity. This can be further validated by calculating the electrostatic interaction of the DMM and TOM coadsorption complex with the zeolite framework. In comparison to the supercage of USY, the isosurface plots at the intersection of ZSM-5 exhibited a larger green region, which represents the electrostatic stabilization effect (Figure 5). This indicates that the maximum included sphere of ZSM-5 can better stabilize the DMM and TOM coadsorption complex through an appropriate electrostatic stabilization effect, which promotes their further conversion into $PODE_n$ [34,35]. The density functional theory calculation provided further evidence that the catalytic activity and product selectivity of various zeolites in converting DMM and TOM to $PODE_n$ were strongly related to the pore mouth size of the zeolite and the interaction between reactants and zeolite channel systems. The adsorption energy and electrostatic interaction of reacting molecules with SUZ-4, beta, and MOR in Table S1 and Figure S9 also proved the above argument.

Table 4. Adsorption energy of DMM, TOM, and their combination in ZSM-5 and USY.

Samples	DMM	TOM	DMM + TOM
ZSM-5	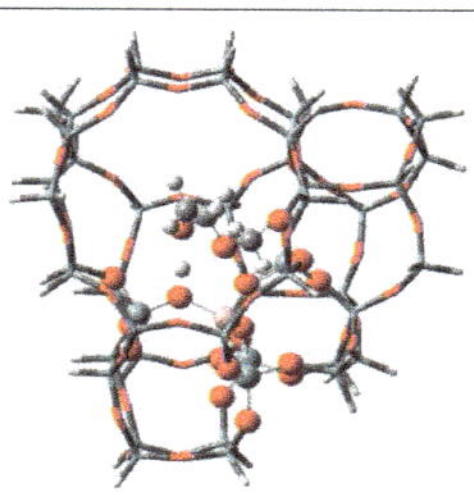 −90 kJ/mol	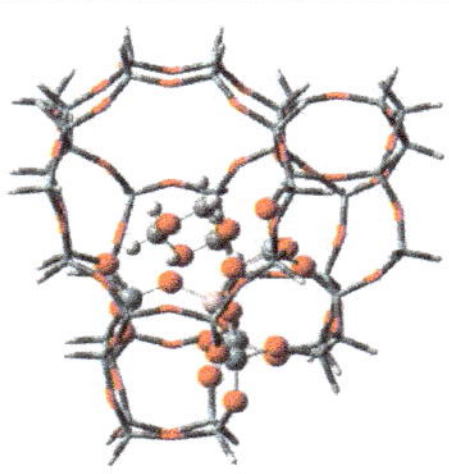 −65 kJ/mol	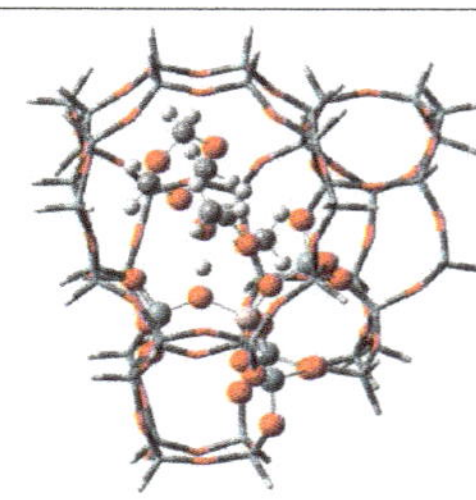 −117 kJ/mol
USY	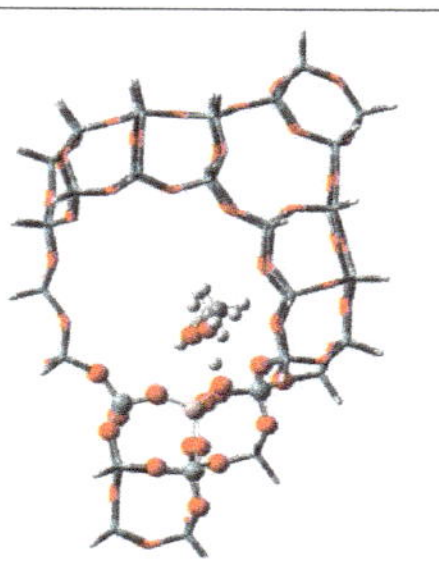 −60 kJ/mol	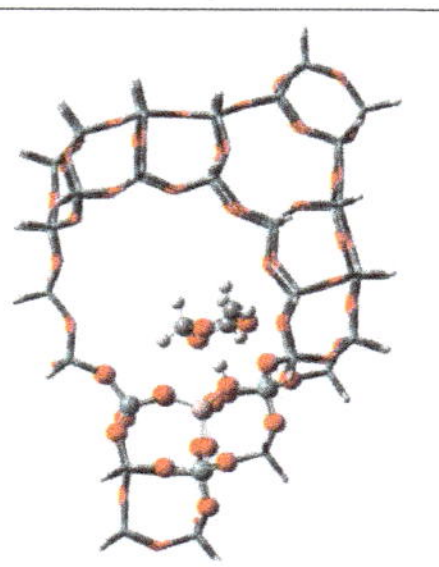 −53 kJ/mol	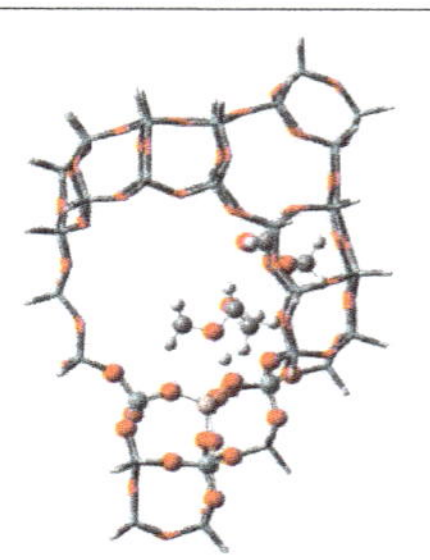 −107 kJ/mol

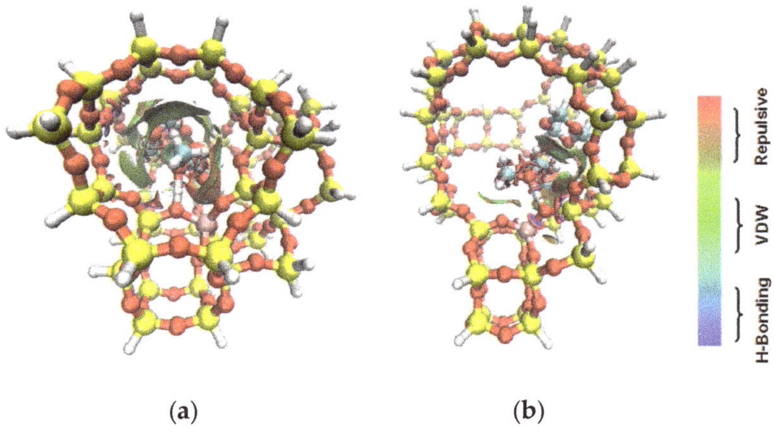

Figure 5. Isosurface plots of the reduced density gradient (s = 0.500 a.u.) for the coadsorption of DMM and TOM in ZSM-5 (**a**) and USY (**b**). The isosurfaces of the reduced density gradient are colored according to the values of the quantity sign (λ2) ρ, as indicated by the RGB scale. VDW: van der Waals interactions.

Another important finding was that increasing reaction temperatures promoted the formation of $PODE_n$ in USY, but rapidly inactivated ZSM-5. This is the result of high temperatures that render intermediate species more active. For USY, the sufficient available space enables a preferential combination of intermediate species from DMM and TOM and restricts side products at high temperatures, resulting in high $PODE_n$ selectivity and very low byproduct selectivity. For ZSM-5 with a Si/Al ratio of 17, the limited space and numerous active acid sites may be more advantageous to TOM decomposition and the subsequent formation of MF by the Tishchenko reaction at high temperatures, while the interaction of DMM with the zeolite and $PODE_n$ synthesis are suppressed. The decomposition of TOM alone over ZSM-5 and USY at 120 °C confirmed this as well. Over HZSM-5, MF was the primary product, whereas over USY, MF almost disappeared in the product, and FA was the dominating product that could take part in the chain propagation forming $PODE_n$ (Figure S10). Increasing the Si/Al ratio of ZSM-5 to 267, i.e., reducing active acid sites (Figure 4), can effectively delay the decline of $PODE_n$ selectivity at high temperatures. This demonstrates that excess active sites and steric constraints of ZSM-5 with a low Si/Al ratio resulted in the enhancement of side reactions at high temperatures.

To sum up, the above facts indicate the obvious synergy between Brønsted MAS and the maximal space available of zeolites for the synthesis of $PODE_n$. Brønsted MAS make TOM and DMM decompose to monomer CH_2O and methoxymethoxy groups (MMZ), the steric constraint of the maximum included sphere with an appropriate size in zeolite channels may help to promote the combination of CH_2O and MMZ to form long-chain $ZO(CH_2O)_nCH_3$, then $ZO(CH_2O)_nCH_3$ react with the methyl-end group to form $PODE_n$ over Brønsted MAS [4]. Besides, increasing reaction temperature can increase intermediate species' activity, so catalytic performance of all studied zeolites is improved gradually before 90 °C. However, too high temperature is not propitious to the formation of $PODE_n$ for low Si/Al ZSM-5, whose excess active acid sites and limited space available may be more adaptable to the conversion of CH_2O species to MF at higher temperature. On the contrary, high temperature can improve the $PODE_n$ selectivity over USY zeolite due to its excessively large space available.

The stability and recyclability are important indexes for potential industrial application, HZSM-5 and USY were further used to evaluate the reusability for the synthesis of $PODE_n$ at 90 °C. As shown in Figure 6, both HZSM-5 and USY catalyst could exhibit excellent stability and reusability; the catalytic activity only displays a very slight decrease after being reused for 10 cycles upon a simple centrifugation separation. At the same time, the catalytic performance of HZSM-5(Si/Al = 17) catalyst was compared with some recently reported catalysts for the synthesis of $PODE_n$ in Table S2. When $PODE_{2-8}$ selectivity and energy consumption are the chief consideration, HZSM-5(Si/Al = 17) showed better catalytic performance than those of many other materials, as illustrated in Table S2.

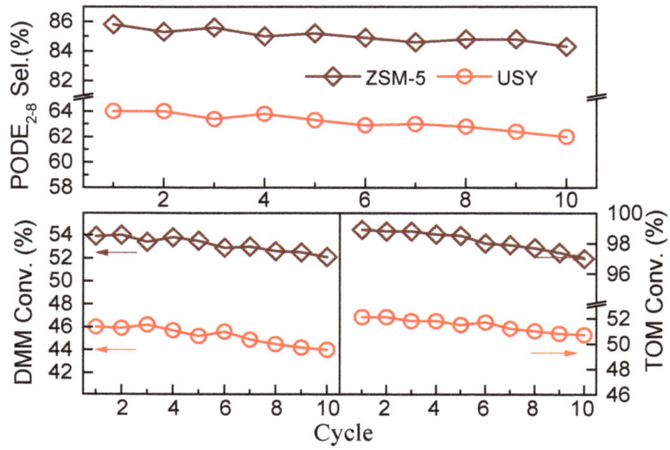

Figure 6. Reusability of the HZSM-5 (Si/Al = 17) and USY (Si/Al = 10) catalyst. The reaction was carried out at 90 °C for 45 min, with a feed DMM/TOM molar ratio of 2 and catalyst amount of 5 wt.%. After each test, the catalyst was reused in the next run following a simple centrifugation separation.

4. Conclusions

The synthesis of $PODE_n$ from DMM and TOM occurred in the zeolite channel. SAPO-34 contained a supercage structure and numerous strong acid sites; however, the 3.8 Å pore mouth diameter prevented TOM from entering, resulting in very poor activity. Comparing with SAPO-34, SUZ-4 exhibited only a slight advantage in catalytic performance due to very limited Brønsted MAS. ZSM-5, MOR, and beta displayed very similar excellent activity at 70–90 °C, which could be ascribed to sufficient Brønsted MAS and effectual steric constraint of the appropriate sizes of the maximum included spheres. Despite its enough acid sites, USY exhibited lower catalytic activity than other zeolites at low temperatures due to its excessively large included spheres, resulting in weak steric constraint for reactive intermediate species. There was a synergy between Brønsted MAS and the maximal space available of zeolites, which interrelated and jointly influenced the synthesis of $PODE_n$ from DMM and TOM. Brønsted MAS could effectively promote the dissociation of DMM and TOM to MM-Z species and CH_2O species; the steric constraint of the maximum included sphere with an appropriate size in zeolite channels could promote the combination of MM-Z and CH_2O to form long-chain $ZO(CH_2O)_nCH_3$ at lower temperatures; $PODE_n$ were ultimately synthesized by $ZO(CH_2O)_nCH_3$ reacting with the methyl-end group over Brønsted MAS.

Reaction temperature also affected the synthesis of the $PODE_n$ by changing the activity of intermediate species. High temperature (>90 °C) may lead to more side reactions and suppress the formation of $PODE_n$ for zeolites with limited space, like low Si/Al ZSM-5 zeolite, but can promote $PODE_n$ selectivity for the zeolites with larger space available, like USY zeolite. Moreover, both HZSM-5 and USY catalysts exhibited excellent stability and reusability.

Altogether, a pore mouth diameter larger than the TOM molecule, a proper amount of Brønsted MAS, and an appropriate maximum included sphere size are necessary conditions to obtain high $PODE_n$ selectivity at low temperatures.

Supplementary Materials: The following are available online at http://www.mdpi.com/2079-4991/9/9/1192/s1, Figure S1: The XRD patterns of various zeolite catalysts; Figure S2: SEM images of various zeolite catalysts; Figure S3: The Py-IR profiles of each zeolite at different temperatures; Figure S4: Effect of the feed DMM/TOM molar ratio on the reactant conversion and $PODE_{2-8}$ selectivity over HZSM-5; Figure S5: Effect of the reaction time on the reactant conversion and $PODE_{2-8}$ selectivity over HZSM-5; Figure S6: Effect of the HZSM-5 catalyst amount on the reactant conversion and product distribution over HZSM-5; Figure S7: SEM images of ZSM-5 zeolites with different particle sizes (I) and its catalytic performance for the synthesis of $PODE_n$ (II); Figure S8: DMM disproportionation over H-ZSM-5 zeolites with a Si/Al ratio of 17 at different temperatures; Figure S9: Isosurface plots of the reduced density gradient (s = 0.500 a.u.) for the coadsorption of DMM and TOM in beta (a) and MOR (b); Figure S10: TOM conversion and product selectivity for TOM decomposition over the ZSM-5 and USY zeolites at 120 °C; Figure S11: Two typical GC chromatograms from Shimadzu GC-2014C and Shimadzu GC-14B; Table S1: Adsorption energy of DMM, TOM, and their combination in SUZ-4, beta, and MOR; Table S2: Comparison of HZSM-5(Si/Al = 17) and some recently reported catalysts for the synthesis of $PODE_n$.

Author Contributions: Experimentation design and the manuscript writing, J.W.; theoretical calculation, S.W.; zeolites synthesis, R.S.; catalytic test and samples characterization, Y.Z. (Yin Zhang) and H.L.; supervision and revision, Y.Z. (Yongxiang Zhao); All authors participated the different stages of the process and approved the final manuscript.

Funding: This research was funded by the financial support of the National Key R&D Program of China (2018YFB0604804) and National Natural Science Foundation of China (21703127, 21802157, U1710221, 21303097, 21503124).

Conflicts of Interest: The authors declare no conflict of interest.

References

1. Wang, D.; Zhu, G.L.; Li, Z.; Xia, C.G. Polyoxymethylene dimethyl ethers as clean diesel additives: Fuel freezing and prediction. *Fuel* **2019**, *237*, 833–839. [CrossRef]
2. Gao, X.J.; Wang, W.F.; Gu, Y.Y.; Zhang, Z.Z.; Zhang, J.F.; Tsubaki, N.; Zhang, Q.D.; Han, Y.Z.; Tan, Y.S. Synthesis of Polyoxymethylene Dimethyl Ethers from Dimethyl Ether Direct Oxidation over Carbon-Based Catalysts. *ChemCatChem* **2018**, *10*, 273–279. [CrossRef]
3. Liu, H.Y.; Wang, Z.; Wang, J.X.; He, X. Performance, combustion and emission characteristics of a diesel engine fueled with polyoxymethylene dimethyl ethers (PODE3-4)/diesel blends. *Energy* **2016**, *97*, 105–112. [CrossRef]
4. Baranowski, C.J.; MBahmanpour, A.; Kröcher, O. Catalytic synthesis of polyoxymethylene dimethyl ethers (OME): A review. *Appl. Catal. B Environ.* **2017**, *217*, 407–420. [CrossRef]
5. Sun, R.Y.; Delidovich, I.; Palkovits, R. Dimethoxymethane as a Cleaner Synthetic Fuel: Synthetic Methods, Catalysts, and Reaction Mechanism. *ACS Catal.* **2019**, *9*, 1298–1318. [CrossRef]
6. Zhao, Y.; Xu, Z.; Chen, H.; Fu, Y.; Shen, J.J. Mechanism of chain propagation for the synthesis of polyoxymethylene dimethyl ethers. *Energy Chem.* **2013**, *22*, 833–836. [CrossRef]
7. Wang, Z.; Liu, H.Y.; Ma, X.; Wang, J.X.; Shuai, S.J.; Reitz, R.D. Homogeneous charge compression ignition (HCCI) combustion of polyoxymethylene dimethyl ethers (PODE). *Fuel* **2016**, *183*, 206–213. [CrossRef]
8. Lautenschutz, L.; Oestreich, D.; Haltenort, P.; Arnold, U.; Dinjus, E.; Sauer, J. Efficient synthesis of oxymethylene dimethyl ethers (OME) from dimethoxymethane and trioxane over zeolites. *Fuel Process. Technol.* **2017**, *165*, 27–33. [CrossRef]
9. Marchionna, M.; Patrini, R. A Process for the Selective Production of Dialkyl-Polyformals. Patent Number. EP1505049A1, 9 February 2005.
10. Oestreich, D.; Lautenschütz, L.; Arnold, U.; Sauer, J. High Purity Oligomeric Oxymethylene Ethers as Diesel Fuels. *Chem. Eng. Sci.* **2017**, *163*, 92–104. [CrossRef]
11. Zheng, Y.; Tang, Q.; Wang, T.; Liao, Y.; Wang, J. Synthesis of a Green Fuel Additive over Cation Resins. *Chem. Eng. Technol.* **2013**, *36*, 1951–1956. [CrossRef]
12. Li, H.J.; Song, H.L.; Chen, L.W.; Xia, C.G. Designed SO_4^{2-}/Fe_2O_3-SiO_2 solid acids for polyoxymethylene dimethyl ethers synthesis: The acid sites control and reaction pathwaysAppl. *Catal. B Environ.* **2015**, *165*, 466–476. [CrossRef]
13. Wang, R.Y.; Wu, Z.W.; Qin, Z.F.; Chen, C.M.; Zhu, H.Q.; Wu, J.B.; Chen, G.; Fan, W.B.; Wang, J.G. Graphene oxide: An effective acid catalyst for the synthesis of polyoxymethylene dimethyl ethers from methanol and trioxymethylene. *Catal. Sci. Technol.* **2016**, *6*, 993–997. [CrossRef]

14. Liu, F.; Wang, T.F.; Zheng, Y.Y.; Wang, J.F. Synergistic effect of Bronsted and Lewis acid sites for the synthesis of polyoxymethylene dimethyl ethers over highly efficient SO_4^{2-}/TiO_2 catalysts. *J. Catal.* **2017**, *355*, 17–25. [CrossRef]
15. Wu, J.B.; Zhu, H.Q.; Wu, Z.W.; Qin, Z.F.; Fan, W.B.; Wang, J.G. High Si/Al ratio HZSM-5 zeolite: An efficient catalyst for the synthesis of polyoxymethylene dimethyl ethers from dimethoxymethane and trioxymethylene. *Green Chem.* **2015**, *17*, 2353–2357. [CrossRef]
16. Fu, W.H.; Liang, X.M.; Zhang, H.D.; Wang, Y.M.; He, M.Y. Shape selectivity extending to ordered supermicroporous aluminosilicates. *Chem. Commun.* **2015**, *51*, 1449–1452. [CrossRef]
17. Zhao, Q.; Wang, H.; Qin, Z.F.; Wu, Z.W.; Fan, W.B.; Wang, J.G. Synthesis of polyoxymethylene dimethyl ethers from methanol and trioxymethylene with molecular sieves as catalysts. *J. Fuel Chem. Technol.* **2011**, *39*, 918–923. [CrossRef]
18. Goncalves, T.J.; Arnold, U.; Plessow, P.N.; Studt, F. Theoretical Investigation of the Acid Catalyzed Formation of Oxymethylene Dimethyl Ethers from Trioxane and Dimethoxymethane. *ACS Catal.* **2017**, *7*, 3615–3621. [CrossRef]
19. Grünert, A.; Losch, P.; Ochoa-Hernández, C.; Schmidt, W.; Schüth, F. Gas-phase synthesis of oxymethylene ethers over Si-rich zeolites. *Green Chem.* **2018**, *20*, 4719–4728. [CrossRef]
20. Xue, Z.Z.; Shang, H.Y.; Zhang, Z.L.; Xiong, C.H.; Lu, C.B.; An, G.J. Efficient Synthesis of Polyoxymethylene Dimethyl Ethers on Al-SBA-15 Catalysts with Different Si/Al Ratios and Pore Sizes. *Energy Fuels* **2017**, *31*, 279–286. [CrossRef]
21. Burger, J.; Ströfer, E.; Hasse, H. Chemical equilibrium and reaction kinetics of the heterogeneously catalyzed formation of poly (oxymethylene) dimethyl ethers from methylal and trioxane. *Ind. Eng. Chem. Res.* **2012**, *51*, 12751–12761. [CrossRef]
22. Baranowski, C.J.; Bahmanpour, A.M.; Héroguel, F.; Luterbacher, J.S.; Kröcher, O. Prominent role of mesopore surface area and external acid sites for the synthesis of polyoxymethylene dimethyl ethers (OME) on a hierarchical H-ZSM-5 zeolite. *Catal. Sci. Technol.* **2019**, *9*, 366–376. [CrossRef]
23. Li, J.F.; Fan, W.B.; Dong, M.; He, Y.; Qin, Z.F.; Wang, J.G. Chemistry Synthesis and MTO Catalytic Performance of SAPO-34. *J. Chin. Univ.* **2011**, *32*, 765–771.
24. Gao, S.; Wang, X.P.; Wang, X.Z.; Bai, Y.X. Green synthesis of SUZ-4 zeolite controllable inmorphology and SiO_2/Al_2O_3 ratio. *Microporous Mesoporous Mater.* **2013**, *174*, 108–116. [CrossRef]
25. Kim, G.J.; Ahn, W.S. Direct synthesis and characterization of high-SiO_2-content mordenites. *Zeolites* **1991**, *11*, 745–750. [CrossRef]
26. Xu, F.; Dong, M.; Gou, W.; Li, J.; Fan, W.; Qin, Z.; Wang, J. Rapid tuning of ZSM-5 crystal size by using polyethylene glycol or colloidal silicalite-1 seed. *Microporous Mesoporous Mater.* **2012**, *163*, 192–200. [CrossRef]
27. García-Trenco, A.; Martínez, A. Direct synthesis of DME from syngas on hybrid CuZnAl/ZSM-5 catalysts: New insights into the role of zeolite acidity. *Appl. Catal. A* **2012**, *411–412*, 170–179. [CrossRef]
28. Ding, X.; Geng, S.; Li, C.; Yang, C.; Wang, G. Effect of acid density of HZSM-5 on the oligomerization of ethylene in FCC dry gas. *J. Nat. Gas Chem.* **2009**, *18*, 156–160. [CrossRef]
29. Li, Y.C.; Wang, H.; Dong, M.; Li, J.F.; Qin, Z.F.; Wang, J.G.; Fan, W.B. Effect of zeolite pore structure on the diffusion and catalytic behaviors in the transalkylation of toluene with 1,2,4-trimethylbenzene. *RSC Adv.* **2015**, *5*, 66301–66310. [CrossRef]
30. Lukyanov, D.B.; Vazhnova, T.; Cherkasov, N.; Casci, J.L.; Birtill, J.J. Insights into Brønsted Acid Sites in the Zeolite Mordenite. *J. Phys. Chem. C* **2014**, *118*, 23918–23929. [CrossRef]
31. Available online: http://www.iza-structure.org/databases (accessed on 18 October 2018).
32. Foster, M.D.; Rivin, I.; Treacy MM, J.; Friedrichs, O.D. A geometric solution to the largest-free-sphere problem in zeolite frameworks. *Microporous Mesoporous Mater.* **2006**, *90*, 32–38. [CrossRef]
33. Celik, F.E.; Kim, T.; Bell, A.T. Effect of zeolite framework type and Si/Al ratio on dimethoxymethane carbonylation. *J. Catal.* **2010**, *270*, 185–195. [CrossRef]

34. Wang, S.; Li, S.Y.; Zhang, L.; Qin, Z.F.; Chen, Y.Y.; Dong, M.; Li, J.F.; Fan, W.B.; Wang, J.G. Mechanistic insights into the catalytic role of various acid sites on ZSM-5 zeolite in the carbonylation of methanol and dimethyl ether. *Catal. Sci. Technol.* **2018**, *8*, 3193–3204. [CrossRef]
35. Wang, S.; Chen, Y.Y.; Qin, Z.F.; Zhao, T.S.; Fan, S.B.; Dong, M.; Li, J.F.; Fan, W.B.; Wang, J.G. Origin and evolution of the initial hydrocarbon pool intermediates in the transition period for the conversion of methanol to olefins over H-ZSM-5 zeolite. *J. Catal.* **2019**, *369*, 382–395. [CrossRef]

nanomaterials

MDPI

Article

Enzymatic Conversion of Oleuropein to Hydroxytyrosol Using Immobilized β-Glucosidase on Porous Carbon Cuboids

Alexandra V. Chatzikonstantinou [1], Elena Gkantzou [1], Eleni Thomou [2], Nikolaos Chalmpes [2], Kyriaki-Marina Lyra [2], Vasiliki G. Kontogianni [3], Konstantinos Spyrou [2], Michaela Patila [1], Dimitrios Gournis [2,*] and Haralambos Stamatis [1,*]

1 Biotechnology Laboratory, Department of Biological Applications and Technologies, University of Ioannina, 45110 Ioannina, Greece
2 Department of Materials Science & Engineering, University of Ioannina, 45110 Ioannina, Greece
3 Section of Organic Chemistry & Biochemistry, Department of Chemistry, University of Ioannina, 45110 Ioannina, Greece
* Correspondence: dgourni@uoi.gr (D.G.); hstamati@uoi.gr (H.S.); Tel.: +30-2651007141 (D.G.); +30-2651007116 (H.S.)

Received: 28 June 2019; Accepted: 11 August 2019; Published: 14 August 2019

Abstract: In the present study, we developed novel β-glucosidase-based nano-biocatalysts for the bioconversion of oleuropein to hydroxytyrosol. Using non-covalent or covalent immobilization approaches, β-glucosidases from almonds and *Thermotoga maritima* were attached for the first time on oxidized and non-oxidized porous carbon cuboids (PCC). Various methods were used for the characterization of the bio-nanoconjugates, such as Fourier transform infrared spectroscopy (FTIR), X-ray photoelectron spectroscopy (XPS), atomic force microscopy (AFM), and fluorescence spectroscopy. The oxidation state of the nano-support and the immobilization procedure play a key role for the immobilization efficiency or the catalytic activity of the immobilized β-glucosidases. The nano-biocatalysts were successfully used for the hydrolysis of oleuropein, which leads to the formation of its bioactive derivative, hydroxytyrosol (up to 2.4 g L^{-1}), which is a phenolic compound with numerous health benefits. The bio-nanoconjugates exhibited high thermal and operational stability (up to 240 h of repeated use), which indicated that they are efficient tools for various bio-transformations.

Keywords: β-glucosidase; carbon cuboids; hydroxytyrosol; oleuropein; bio-catalysis; nano-biocatalyst

---

1. Introduction

Bio-catalysis has rapidly gained ground in almost every catalytic process due to its advantages, such as selectivity (region-, chemo-, stereo-) and low environmental impact when compared to traditional synthetic methodologies. Heterogeneous bio-catalysis refers to enzymes in a water-insoluble form and is preferred lately for industrial production processes, mainly for the easy separation of the products, but also for the ability to reuse the biocatalyst for multiple reaction cycles [1]. Heterogeneous biocatalysts are enzymes immobilized in various solid materials such as polymers, silica, nanomaterials, etc. [2].

The benefits of enzyme immobilization, along with using nanomaterials as matrices, have gained interest during the last decade [3]. Carbon-based nanomaterials have dominated the world of nano-immobilization, since they combine both effectiveness and biocompatibility [4,5]. Researchers have managed a wide range of carbon nanomaterials, regarding both shapes and sizes, with each one competing each other for its benefiting effect on enzymatic performance. It is well known, that the structural characteristics of the nanomaterials, such as size, shape, porosity, and surface chemistry, can affect the immobilization and the catalytic behavior of the immobilized enzyme. Carbon porous nanomaterials have long attracted the attention of enzyme supports, due to their important features, regarding enzyme immobilization, such as the increased surface area and the pore volume they provide in order to achieve higher protein loading, while they exhibit no spatial restrictions upon enzyme molecules [6–8]. Porosity also seems to influence the activity of the biocatalyst since it can facilitate substrate accessibility to the enzyme [9,10]. Functionalized carbon-based nanomaterials also excel for their unique properties that functional groups provide them with. Different kinds of functionalization modify the properties of the nanomaterials, which affects their interaction with proteins, and, thus, the catalytic characteristics of the immobilized enzymes [11,12]. Several studies have shown that the functionalization of carbon-based nanomaterials enhances not only the enzyme loading and the catalytic activity, but also the operational and thermal stability of the biocatalysts [13,14].

A brand-new member of the carbon family, which combines porosity and capability to accept functional groups on its surface, has recently been reported as a high performance material [15]. This novel nanostructure is called porous carbon cuboids (PCCs) and combines a series of intriguing properties, such as light weight, unusual ultra-hydrophilic behavior, great stability, surface heterogeneity, and a very high hierarchical porosity (estimated approximately 800–900 m^2 gr^{-1}) [15]. In contrast with graphene, PCC do not need to pass through the oxidation state due to the high number of functional groups they possess on the surface, which renders them highly hydrophilic. PCC show a significant number of N:C and O:C active sites, which, in combination with the narrow micropore size distribution, constitute a very promising sorbent. The advantage that they present over other carbon materials is the fact that they are stable up to 400 °C, whereas most of them cannot be functional in temperatures further than 300 °C. By further treatment of the PCCs with strong oxidizing agents (employing a modified Staudenmaier's method), multiple oxygen functionalities (such as carboxyl, hydroxyl, and epoxy) can be introduced, which increases the number of active sites [16–20]. These oxidized PCCs, as well as the pristine ones, have the ability to interact with bacteria and algae, as recently shown [16].

In the present study, pristine PCCs and surface oxidized (with -COOH and -OH groups) analogues (PCCox) were used, for the first time, as nano-supports for the covalent and non-covalent immobilization of two β-glucosidases (from almond and *Thermotoga maritima*) that are widely used in various biocatalytic processes with industrial interest [21,22]. These novel nano-biocatalysts were tested for their ability to catalyze the bioconversion of oleuropein (OLE) to hydroxytyrosol [3,4-dihydroxyphenylethanol (HT)] (Figure 1). OLE is a hydrophilic phenolic antioxidant found in all parts of the olive tree, with high concentrations in the dry leaf extract and also abundant in olive mill wastewaters. HT is the main degradation product of OLE, which is considered to be one of the strongest antioxidants in nature [23] with superior biological activities than the parent compound OLE [24,25]. HT is used as a food supplement and is used in the cosmetic industry. In addition, it is considered to be a potent drug agent and a food additive due to its anti-inflammatory, anticarcinogenic, neuroprotective, and antiapoptotic activity [23]. However, HT is found in low concentration in nature, has low extraction yields from natural sources, and is difficult to chemically synthesize. Therefore, the production of pure HT is a high cost procedure [26]. Furthermore, the development of biotechnological approaches for the synthesis of HT is of great interest. The synthesis of HT through the conversion of OLE from olive leaves or olive mill wastewaters could be an optimal direction for a cost-effective production of HT [23]. Various hydrolytic enzymes have been employed for the conversion of OLE to HT such as esterases, lipases, xylanases, cellulases, and hemicellulases [27]. Natural or recombinant β-glucosidases have been proved to be the key enzymes for OLE hydrolysis and have been used either as free or immobilized enzymes [27–31].

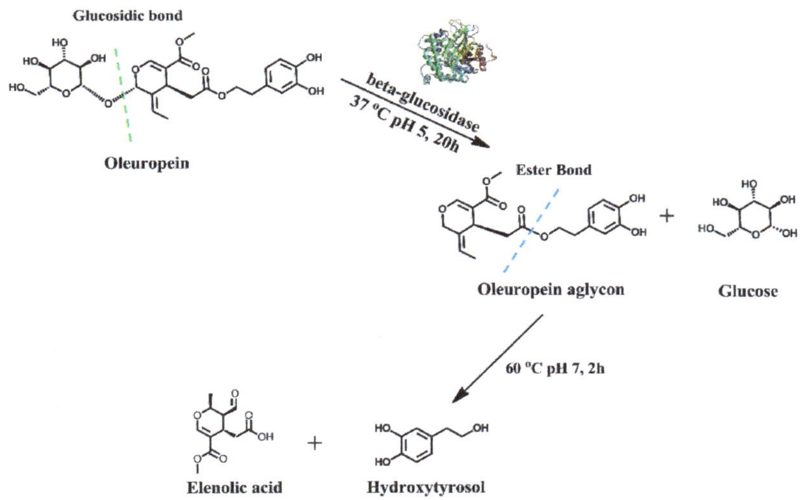

Figure 1. Transformation of oleuropein to hydroxytyrosol.

A combination of spectroscopic, microscopic, and biochemical techniques was applied for the characterization of these novel bio-nanoconjugates with respect to their catalytic behavior. Moreover, the effect of the surface chemistry of PCCs and the immobilization approaches that were used on the catalytic activity as well as thermal and operational stability of the immobilized enzymes was investigated.

2. Materials and Methods

2.1. Materials

β-Glucosidase from almonds (Albgl) 9.5 U mg^{-1} was purchased from Sigma-Aldrich (St. Louis, MO, USA) and was used with no further purification. N′-ethylcarbodiimide hydrochloride (EDC), N-Hydroxysuccinimide (NHS), *p*-nitrophenol (*p*-NP), and 4-Nitrophenyl *β*-*D*-glucopyranoside (*p*-NPG) were purchased from Sigma-Aldrich (St. Louis, MO, USA), *β*-glucosidase from *Thermotoga maritima* (Tmbgl) 85 U mg^{-1} was purchased from Megazyme (Chicago, IL, USA) and was used with no further purification. Oleuropein and hydroxytyrosol were purchased from Extrasynthese (Lyon, France). 4,4′-Bipyridine (98%) and Potassium chlorate ($KClO_3$, 99+%) were purchased from Alfa-Aesar (Kandel, Germany). Ethanol (absolute, for analysis) was purchased from Merck (KGaA Darmstadt, Germany). Nitric acid (HNO_3, 65%) was purchased from Chem-Lab (Zedelgem, Belgium), and Sulphuric acid (H_2SO_4, 96%) was purchased from Panreac (Castellar del Vallès, Spain).

2.2. Synthesis of Porous Carbon Cuboids (PCC) and Oxidized PPC (PCCox)

The same procedure that has been described in our previous work was followed for the synthesis of PCC [16]. Two solutions were prepared. The first one consisted of 1 g of Pluronic dissolved in 100 mL of 0.1 M 4,4′-bipyridine in water−ethanol (volume ratio of 1:17) and the second one was an aqueous solution of copper (II) chloride (900 mL, 5.6 mM $CuCl_2{\cdot}2H_2O$). The first solution was then poured rapidly into the second one under vigorous stirring. The formed products were retrieved through centrifugation, washed three times with water, and air dried. The product was then pyrolyzed under argon flow (500 °C for 2 h) and the copper species were leached away (stirred in a 4 M HNO_3 aqueous solution for 24 h). The final black powder was collected by air drying after being washed with water until pH was ~5.0.

Likewise, the PCC was treated with strong oxidizing agents by employing a modified Staudenmaier's method, in order to introduce more active sites (PCCox) [16–20]. Additionally, 70 mg of PCC were added to a mixture of H_2SO_4 and HNO_3 (1.4 mL and 2.8 mL, respectively) while keeping the temperature at ~0 °C in an ice-water bath. Under constant stirring and cooling, small quantities of $KClO_3$ (0.7 g in total) were added slowly to the mixture. The reaction was terminated after 18 h by pouring the mixture into distilled water and the final product was washed until pH~6.0 and was dried at room temperature.

2.3. Covalent Immobilization of β-Glucosidase on PCC and PCCox

Albgl and Tmbgl were covalently immobilized on PCC and PCCox via cross-linking agents such as EDC and NHS, which can link the -COOH and -OH groups of nanomaterials with the free amino groups of the enzyme [32,33]. In a typical procedure, 4 mg of nanomaterial were dispersed in 5 mL H_2O and 1 mL HEPES (*N*-(2-Hydroxyethyl) piperazine-*N*′-(2-ethanesulfonic acid)) solution (pH 7.0, 50 mM) in an ultrasonic bath for 30 min. After the dispersion, 2.3 mL of a 50 mg mL^{-1} NHS aqueous solution and 1.2 mL of a 10 mg mL^{-1} EDC aqueous solution were added and the mixture was incubated under stirring for 30 min at 30 °C. The modified nanomaterial was then separated by centrifugation at 4000 rpm for 10 min and washed with HEPES solution in triplicate to remove the excess of EDC and NHS. The activated nanomaterial was re-dispersed in 6 mL HEPES solution 50 mM at the optimum pH values (pH 5.0 and pH 6.5 for Albgl and Tmbgl, respectively). Then, 0.3 mg (estimated by Bradford assay) was added and the mixture was incubated under stirring at 30 °C for 1 h. The bio-nanoconjugates were separated by centrifugation at 4000 rpm for 10 min and washed with the proper HEPES solution in triplicate to remove the excess of enzyme and then left to dry in silica at 4 °C. The nano-biocatalysts were labeled as PCC-Albgl-cov, PCCox-Albgl-cov, PCC-Tmbgl-cov, and PCCox-Tmbgl-cov.

2.4. Non-Covalent Immobilization of β-Glucosidase on PCC and PCCox

Albgl and Tmbgl were attached via physical absorption on PCC and PCCox. Additionally, 4 mg of nanomaterial were dispersed in 6 mL HEPES solution 50 mM, at the optimum pH values (pH 5.0 and pH 6.5 for Albgl and Tmbgl, respectively), in an ultrasonic bath for 30 min. Furthermore, 0.3 mg of β-glucosidase were added and the mixture was incubated under stirring at 30 °C for 1 h. The bio-nanoconjugates were separated by centrifugation at 4000 rpm for 10 min and washed with the proper HEPES solution in triplicate to remove the excess of the enzyme and then left to dry in silica at 4 °C. The nano-biocatalysts were labeled as PCC-Albgl-nc, PCCox-Albgl-nc, PCC-Tmbgl-nc, and PCCox-Tmbgl- nc.

2.5. Determination of the Immobilization Yield

The Bradford assay was used to determine the immobilization yield by estimating the amount of the protein in the supernatant before and after the immobilization [34]. The amount of immobilized β-glucosidase on PCC and PCCox was calculated as the difference of the protein concentration in the supernatant before and after the immobilization. All experiments were carried out in duplicate.

2.6. Determination of the Hydrolyitc Activity of Immobilized β-Glucosidase

The hydrolytic activity of β-glucosidases was determined spectrophotometrically by estimating the *p*-NP released from the hydrolysis of *p*-NPG at 410 nm, as described before [35]. The catalytic assay was carried out in citrate-phosphate buffer solution 100 mM, pH 5.0, and pH 6.5 for Albgl and Tmbgl, respectively, at 50 °C for 10 min. The reaction was started by adding an appropriate amount of immobilized β-glucosidase (0.1 or 0.01 mg mL^{-1}) in a *p*-NPG solution (2 mM). The reaction was stopped by adding 0.5 mL of a 10% w/v Na_2CO_3 solution and the reaction product (*p*-NP) was measured at 410 nm. The amount of the released *p*-NP was quantified using a *p*-NP standard curve. One β-glucosidase unit (U) was defined as the μmol min^{-1} of *p*-NP that results from 1 mg of immobilized enzyme under the above-defined conditions.

2.7. Thermal Stability of Free and Immobilized β-Glucosidases

Thermal stability studies of β-glucosidases were performed by incubating free or immobilized enzymes in citrate-phosphate buffer 100 mM, pH 5.0, and pH 6.5 for Albgl and Tmbgl, respectively, at 60 °C for 24 h. The amount of the enzyme was 0.01 mg mL^{-1} or 0.01 μL mL^{-1} of free Albgl and Tmbgl, respectively, and 1 mg mL^{-1} of immobilized β-glucosidase, in all cases. The remaining hydrolytic activity of β-glucosidase was determined, at different time intervals, through the hydrolysis of *p*-NPG, as described before. All experiments were performed in triplicate.

2.8. Hydrolysis of Oleuropein to Hydroxytyrosol

The enzymatic hydrolysis of OLE was performed in citrate-phosphate buffer 100 mM, pH 5.0, and pH 6.5 for Albgl and Tmbgl, respectively. In addition, 1 mg of a nano-biocatalyst was dispersed in 2 mg mL^{-1} solution of OLE (3.7 mM), which was followed by 24 h incubation at 37 °C under shaking (750 rpm). After the enzymatic hydrolysis of OLE (step 1) (Figure 1), a second non-enzymatic step followed to give rise to HT. In step 2, the reaction mixture was incubated at 60 °C, pH 7.0 for 2 h under shaking (750 rpm).

2.9. High Performance Liquid Chromatography (HPLC) Analysis

The identification and quantification of OLE and HT were performed by high performance liquid chromatography (HPLC) (Shimadzu, Tokyo, Japan) using a μBondapack C18 column, particle size 10 μm, length 300 mm, diameter 3.9 mm, and a diode array UV detector. The mobile phase was acetonitrile (A) and 0.1% acetic acid in water (B) of 20–80%. The elution conditions applied were: 0–30 min, 50–50% (v/v) A:B, 30–35 min, 50–50% A:B and 35–40 min, and 20–80% A:B. The elution was performed at 27 °C with a flow rate of 1 mL min^{-1} and the samples were detected at 280 nm. The quantification/characterization of OLE and HT was based on standard samples and calibration curves at the same conditions.

2.10. Reusability Studies of Immobilized β-Glucosidases

The reusability of immobilized β-glucosidases was estimated by performing consecutive operating cycles of the hydrolysis of OLE, as described before. Each reaction cycle was carried out for 24 h at 37 °C under shaking (750 rpm). After each run, the immobilized enzyme was separated from the reaction mixture with centrifugation and washed thoroughly with citrate-phosphate buffer solution 100 mM, pH 5.0, and pH 6.5 for Albgl and Tmbgl, respectively, and then it was re-used. All experiments were performed in triplicate.

2.11. Fourier-Transform Infrared Spectroscopy (FTIR)

An FTIR-8400 infrared spectrometer (Shimadzu, Tokyo, Japan) equipped with a deuterated triglycine sulfate (DTGS) detector was used for the Fourier-transform infrared spectroscopy (FTIR) analysis. All spectra were recorded in the range of 400 to 4000 cm^{-1} and were an average of 32 scans. All samples were in the form of KBr pellets containing ca. 2 wt% of the enzyme, the nanomaterials (PCC and PCCox), or the nano-biocatalysts. Equation (1) was used to calculate the correlation coefficient *r* in order to determine the similarity between two FTIR spectra [36–39].

$$r = \sum x_i y_i / \sqrt{\sum x_i^2 \sum y_i^2}, \tag{1}$$

where x and y are the spectral absorbance values of free and immobilized enzyme, respectively, at the ith frequency position. For the calculation, the absorbance values of the spectra after smoothing in the region of 1600 to 1700 cm^{-1} (amide region I) were used. The correlation coefficient r value will be equal to one ($r = 1$) for identical spectra.

2.12. Fluorescence Spectroscopy

A luminescence spectrofluorometer Jasco-8300 (Tokyo, Japan) was used for all fluorescence measurements using a solid sample holder. Immobilized β-glucosidase was used in aqueous solutions at concentration of 1 mg mL^{-1}, while free Albgl and Tmbgl at concentration of 0.13 mg mL^{-1}. Samples were deposited onto silicon wafers (P/Bor, single side polished) from aqueous solutions by drop casting. The fluorescence emission spectra were recorded from 300 to 400 nm after exciting at 280 nm, with a scan speed of 100 nm min^{-1} at room temperature. Slit widths with a nominal band pass of 5 nm were used for both excitation and emission ray.

2.13. X-ray Photoelectron Spectroscopy

X-ray photoelectron spectroscopy (XPS) measurements were performed in ultra-high vacuum at a base pressure of 2×10^{-10} mbar with a SPECS GmbH spectrometer equipped with a monochromatic MgKa source (hv = 1253.6 eV) and a Phoibos-100 hemispherical analyzer (Berlin, Germany). The spectra were collected in normal emission and energy resolution was set to 1.16 eV to minimize measuring time. All binding energies were referenced to the C1s core level at 284.6 eV [40]. Spectral analysis included a Shirley background subtraction and a peak deconvolution employing mixed Gaussian-Lorentzian functions, in a least square curve-fitting program (WinSpec) developed at the Laboratoire Interdisciplinaire de Spectroscopie Electronique, University of Namur, Belgium.

2.14. Atomic Force Microscopy

Atomic force microscopy (AFM) images were collected in tapping mode with a Bruker Multimode 3D Nanoscope (Ted Pella Inc., Redding, CA, USA), using a microfabricated silicon cantilever type TAP-300G, with a tip radius <10 nm and a force constant of ~20–75 N m^{-1}.

2.15. Raman Spectroscopy

Raman spectra were collected with a Micro-Raman system RM1000 RENISHAW (RENISHAW, Old Town, UK) using a laser excitation line at 532 nm (laser diode) in the range of 1100 to 1800 cm^{-1}. Raman scattering was collected by means of an optical microscope equipped with 50× and 100× lenses.

3. Results and Discussion

In the present work, we developed robust nano-biocatalysts as efficient tools in various biocatalytic processes with industrial interest. More specific, we investigated the immobilization of β-glucosidases on PCCs and functionalized PCCs with -COOH and -OH groups. The enzymes were immobilized with both covalent and non-covalent approaches. Covalent immobilization was carried out through EDC/NHS linkers while non-covalent immobilization was performed through physical adsorption. This way, we could investigate how the immobilization approach or the surface chemistry of the nanomaterial affects the immobilization efficiency, as well as the catalytic behavior of the immobilized β-glucosidases.

3.1. Characterization of PCC and PCCox

Raman, FTIR, and XPS spectroscopy were employed to characterize both the pristine and oxidized PCC. The infrared spectra of PCC and PCCox are depicted in Figure 2a. In the infrared (IR) spectra of the PCC, the peaks between 2843 and 2931 cm^{-1} are due to the stretching vibrations of CH_2 groups, while the broad band centered to 3431 cm^{-1} is attributed to the adsorbed H_2O deformation, which is indicative of the hydrophilic nature of the carbon material. The weak band of the hydroxyl stretching vibrations of the C-OH groups, expected around 3550 cm^{-1} [41], is not visible since it is superimposed in this broad. Moreover, the band at 1402 cm^{-1} is attributed to the aromatic C=C functional group, while the main peak at 1623 cm^{-1} is due to C=O stretching vibrations of the carboxyl groups [42]. The latter is more pronounced and intense in the case of PCCox spectrum due to a higher degree of oxidation. In the same way, the band at 1256 cm^{-1}, due to vibrations of the epoxy groups (COC) [43–45], is more distinct in the PCCox sample, compared to the pristine one, as a result of the extensive oxidation process.

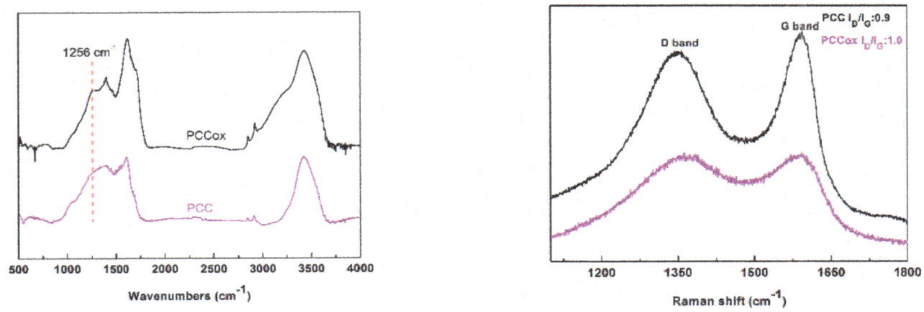

(a) (b)

Figure 2. (**a**) Fourier-transform infrared spectroscopy (FTIR) spectra of the porous carbon cuboids (PCC) and oxidized porous carbon cuboids (PCCox). (**b**) Raman spectra of the PCC and PCCox.

The Raman spectra of the porous carbon cuboids before and after oxidation are shown in Figure 2b. The two characteristic graphitic peaks, at 1350–1370 and 1593 cm^{-1} corresponding to graphitic D and G bands, respectively, were observed in both Raman spectra of the synthesized nanomaterials. The D band is associated with an sp^3-hybridized carbon, whereas the G-band is associated with an sp^2-hybridized carbon atoms. The intensities ratio I_D/I_G express the degree of disorder of the carbon lattice. After oxidation, the ratio was lightly increased to 1.0 (from 0.9 in the pristine PCC), which indicates a change in hybridization of the carbon atoms from sp^2 to sp^3 due to the increase of the population of the oxygen-containing groups, as the result of the oxidation process [46].

The successful oxidation of the PCC was further supported by XPS spectroscopy. The results of pristine PCC and oxidized PCC (PCCox) have been reported in our previous work [16]. According to this work, pristine PCC possess oxygen functionalities, apart from the main C-C frame, such as C-O, C=O, and C(O)O. After oxidation, a significant decrease to the C-C bond (from 51.6% for PCC to 34% for PCCox) can be deduced from the insertion of multiple oxygen moieties to the main body of PCC. Most of the oxygen functionalities increased in ratio (such as C-OH and C(O)O), while the insertion of an additional peak is observed, derived from the epoxy groups due to oxidation treatment.

Representative AFM images of the PCC and PCCox are presented in Figure 3, which confirms the formation of the cubic structure. In addition, the topographic image of an isolated cuboid, as revealed from the cross-sectional analysis, is observed in Figure 3d with an average thickness of 8.51 nm. Accordingly, from the AFM images of PCCox, it is clearly observed that the oxidation process influenced the cubic shape of the carbon nanostructures, as well as the thickness of the PCCox, which is decreased, as shown in Figure 3f. All these values are in agreement with our previous work [16]. It is interesting to note that, while isolated cubes are observed frequently in the surface of the Si-wafer, there are areas where PCCs tend to get organized into ring structures, by integrating monomeric cuboids (Figure 3a,b,d). This behavior could be due to intramolecular interactions between adjacent cuboid nanoparticles in water, where the formation of hydrogen bonding (via carboxyl and/or hydroxyl groups) occurs between neighboring cuboids, which results in the formation of a 2D ring-like supramolecular structure on the surface of a silicon wafer. In fact, when pH was raised to 11.0 (from 5.6 that was initially), due to deprotonation of the carboxyl (or hydroxyl) groups, intramolecular hydrogen bonding was no longer possible and no ordered structure was observed on the surface. The presence of water seems to be crucial for the formation of these ring-like structures, since, when experiments were performed in ethanol (instead of water), no similar ring-like formations were observed. On the other hand, oxidized PCCs, due to the greater number of oxygen functionalities on their surface, are more hydrophilic than pristine PCCs and prefer to be dispersible and isolated in the aqueous medium. Thus, upon deposition, the majority do not tend to aggregate into ring structures (Figure 3e,f).

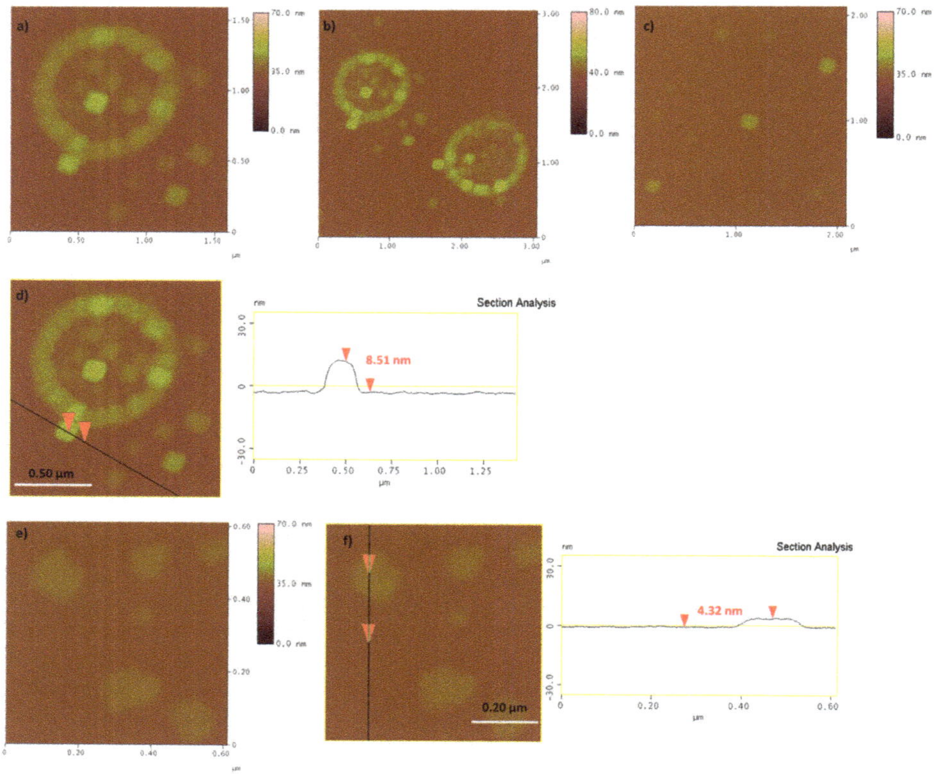

Figure 3. Atomic force microscopy (AFM) height images and cross section analysis of PCC (**a–d** images) and PCCox (**e**,**f** images).

3.2. Immobilization Efficiency and Activity of Immobilized β-Glucosidase

In the present work, PCCs and functionalized PCCs with multiple oxygen functionalities (PCCox) were used, for the first time, as nano-supports for enzyme immobilization. It is well known that the functionalization of carbon-based nanomaterials plays a key role in the development of interactions between the nanomaterial and the enzyme molecules, which affects the catalytic behavior of the latter [4,12,47]. The immobilization of Albgl and Tmbgl was carried out through physical adsorption and covalent immobilization. Physical adsorption of the enzyme onto carbon-based nanomaterials is based on hydrophobic interactions, electrostatic and van der Walls forces, and hydrogen bonding [12,48]. Covalent attachment of the enzyme onto the -OH and -COOH groups of these nanomaterials was carried out by using EDC and NHS as cross-linkers. In this case, stable amide bonds are formed between the enzyme and the nanomaterial [48], even though the co-existence of physical adsorption of the enzyme cannot be excluded.

Table 1 presents the immobilization yield of Albgl and Tmbgl on PCC and PCCox. In all cases, the mass ration of the enzyme to the nanomaterial was 0.1. Tmbgl was successfully immobilized, for the first time onto nanomaterials. The immobilization efficiency of Tmbgl does not seem to depend on the immobilization procedure or the nano-support. It should be noted that, in a range of pH 5–7, no significant differences in the immobilization yields were observed. As it can be seen in Table 1, high immobilization yields are achieved (>90%) in all cases. β-Glucosidase from *Thermotoga maritima* was previously successfully immobilized on chitin beads and was used for the hydrolysis of lactose [49]. On the other hand, the immobilization of Albgl on PCC and PCCox is also successful but the immobilization yield seems to depend on the immobilization procedure and the nano-support that was used. The highest immobilization yield (90%) is observed when PCCox is used as a nano-support for the covalent immobilization of the enzyme, which points out the importance of the functionalization of PCCs. Previous works have shown that covalent immobilization on surface O-functionalities can increase the enzyme loading [11,50]. As far as it concerns non-covalent immobilization, a higher immobilization yield is obtained in the case of PCCs. The presence of oxygen functional groups can lead to lower immobilization yields on carbon-based nanomaterials, as reported elsewhere [47]. Similar, and, in some cases, lower, immobilization efficiencies have been previously reported for β-glucosidase immobilized on various nanomaterials, such as single-walled and multi-walled carbon nanotubes and hybrid magnetic graphene oxide nanoparticles [35,51,52].

Table 1. The immobilization yield (%) and enzymatic activity of β-glucosidase on PCC and PCCox after covalent and non-covalent immobilization (the standard deviation was less than 5% in all cases).

	Immobilization Yield %–(Activity U mg^{-1} Enzyme)			
	Albgl		**Tmbgl**	
	Covalent	Non-covalent	Covalent	Non-covalent
PCC	72–(0.8)	80–(0.5)	95–(24)	94–(18)
PCCox	90–(5)	62–(9)	93–(35)	94–(37)
free	9.5		45	

The hydrolytic activity of the bio-nanoconjugates was investigated and the initial reaction rates are presented in Table 1. The catalytic activity of free β-glucosidases is higher compared to the immobilized enzymes. This reduction in activity is in accordance with that previously reported for β-glucosidase and other enzymes, and could be attributed either to conformational changes in the enzyme molecule upon immobilization or to mass transfer effects, which can reduce the catalytic efficiency of the immobilized enzyme [35,53]. It is interesting to note that the catalytic activity of the immobilized β-glucosidase does not seem to be correlated with the enzyme loading, but it seems to depend on the nature of the nano-support. As can be seen in Table 1 when PCCox is used as a nano-support, higher hydrolytic activity is achieved in all cases. The reduced activity that is observed in the case of PCCs could be attributed to the interaction of enzymes with the hydrophobic surface of the nanomaterials, which could lead to undesirable conformational changes in the protein molecules, and, hence, loss of their catalytic activity. Similar behavior was observed for the immobilization of lipases on carbon nanotubes and graphene-based nanomaterials [4]. Moreover, it is expected that a significant part of the enzyme molecules is mainly adsorbed in the macro-pores of the PCCs, which causes restricted diffusion of the substrate on the active sites of immobilized enzyme molecules [6]. On the other hand, the oxidation of PCC is expected to reduce the porosity of the nanomaterials, which prevents the adsorption of the enzymes on the pores of the nanomaterials. Furthermore, it is expected that the enzymes are attached to the oxygen functional groups of the PCCox nanomaterials. The presence of these oxygen functional groups could create a more hydrophilic microenvironment around immobilized enzyme molecules that facilitates the diffusion of the substrate to the active site of the enzyme, and, thus, increases its catalytic activity [54].

3.3. *Characterization of Bio-Nanoconjugates*

To confirm the presence of β-glucosidase on PCC and PCCox, FTIR spectroscopy was used. Figure 4 illustrates the FTIR spectra of free and immobilized Albgl and Tmbgl. The spectrum of free Albgl shows a typical protein spectrum that presents two absorption bands at 1540 cm^{-1} and 1650 cm^{-1} [55], associated with the amide bond due to the C=O stretching vibrations, and is directly related to the backbone conformation. Immobilized Albgl also depicts two bands at 1540 cm^{-1} and 1650 cm^{-1}, which suggests the successful conjugation of the enzyme onto the PCC and PCCox. Similarly, the spectrum of free Tmbgl presents three absorption bands at 1378 cm^{-1}, 1343 cm^{-1}, and 1639 cm^{-1} associated with the amide bond region. Immobilized Tmbgl depicts a band at 1068 cm^{-1} and a band at 1378 cm^{-1}, which is slightly transposed when compared to the free enzyme. These peaks indicate the successful immobilization of the enzyme onto the PCC and PCCox. FTIR analysis was also employed to investigate the conformational changes of the enzyme upon immobilization onto PCC and PCCox compared to the structure of the free enzyme. To evaluate the differences between the FTIR spectra of free and immobilized β-glucosidase, the correlation coefficient, r, was estimated [38,39,56]. According to the results, the correlation coefficient r for immobilized Albgl and Tmbgl is 0.99 and 0.98, respectively, which indicates that the structure of the immobilized enzyme is not significantly altered.

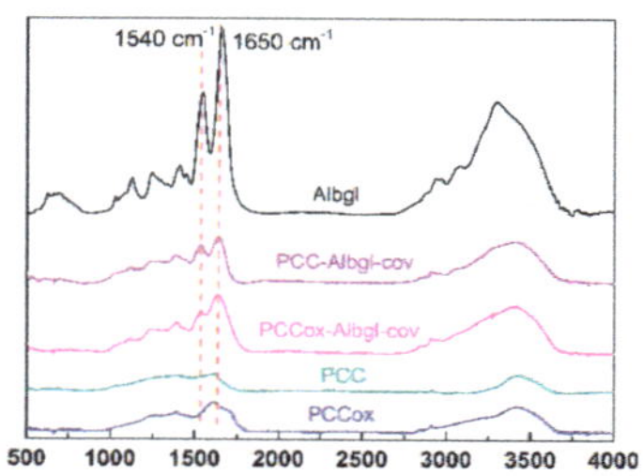

(a)

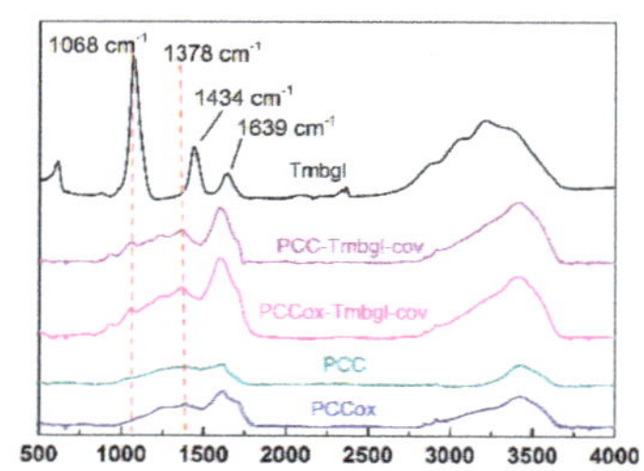

(b)

Figure 4. FTIR spectra of (**a**) PCC-Albgl-cov and PCCox-Albgl-cov compared to those of free Albgl and bulk PCC and PCCox. (**b**) PCC-Tmbgl-cov and PCCox-Tmbgl-cov compared to those of Tmbgl and bulk PCC and PCCox.

Fluorescence spectroscopy has been used before to confirm the presence of the enzyme on solid material [57,58]. Herein, the fluorescence emission spectra of free and immobilized enzymes were recorded in order to further confirm the presence of β-glucosidase on PCC and PCCox (Figure 5). Excitation was set at 280 nm and the emission intensity was recorded in the range of 300 to 400 nm. The maximum emission wavelength of free β-glucosidase was found at 320 nm for both Albgl and Tmbgl. The fluorescence spectra of bulk nanomaterials PCC and PCCox were also recorded at the same conditions and no emission was observed in the range of 300 to 400 nm. As can be seen in Figure 5, in all cases, the fluorescence spectra of bio-nanoconjugates reveal a maximum emission at 320 nm, which indicates the presence of β-glucosidase and, therefore, the successful immobilization of the enzyme.

X-ray photoelectron spectroscopy of bio-nanoconjugates was employed to the porous cuboids and porous cuboids after the oxidation process in order to distinguish the type of interaction (covalent or non-covalent) between the enzymes and the PCC matrix. From the carbon 1s photoelectron spectra of the PCC-Tmbgl-cov (Figure 6a), five contributions are deduced from the C=C/C-H bonds at 284.6 eV (33.7%), the C-O/C-N bonds from the C-O functional groups of PCC, and the C-N and C-O bonds from the enzyme representing 40.1% of the whole carbon spectra. At 287.2 eV, the contribution of C=O bonds (14.1%) is detected, while a very unique peak that attests to the successful covalent bonding between the Tmbgl and PCC is observed at 288.3 eV due to the creation of the amide groups [59] representing 9.0% of the carbon amount. Lastly, a last fitted peak at 289.2 eV is due to the carboxyl groups. This peak represents just 3.1% and this is because carboxyl groups participate in the covalent bond to create amide groups. Accordingly, from the carbon 1s photoelectron spectra of PCC-Tmbgl-nc (Figure 6b), four contributions are displayed due to C=C/C-H located at 284.6 eV covering 48.4%

of the carbon spectra, C-O/C-N groups centered at 285.8 eV (32.8%), carbonyl groups at 287.4 eV (12.7%), and, lastly, a last fitted peak at 289.0 eV due to carboxyl groups representing 6.0% of the whole carbon amount. No amide peak is observed, while the carboxyl group peak is much higher than in the case of PCC-Tmbgl-cov, which is a sign of non-covalent interaction between the enzyme and the PCC nanomaterials.

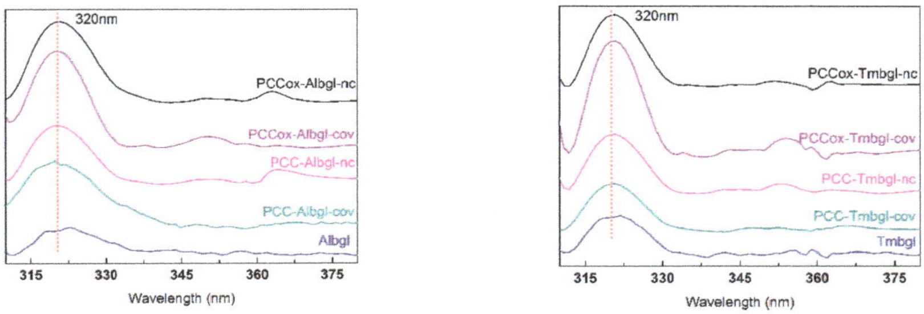

(a) (b)

Figure 5. Fluorescence spectra of (**a**) free and immobilized Albgl on PCC and PCCox. (**b**) Free and immobilized Tmbgl on PCC and PCCox.

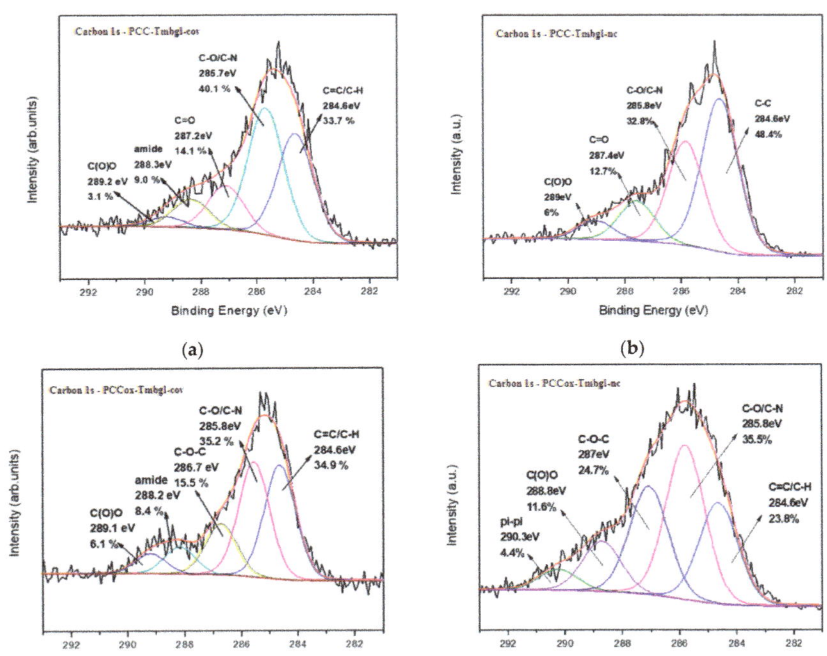

(a) (b)

(c) (d)

Figure 6. C1s photoelectron spectra of (**a**) PCC-Tmbgl-cov, (**b**) PCC-Tmbgl-nc, (**c**) PCCox-Tmbgl-cov, and (**d**) PCCox-Tmbgl-nc.

XPS spectra of PCCox-Tmbgl-cov are displayed in Figure 6c. Fitting on the C1s photoelectron peak reveals five different contributions derived from C=C/C-N at 284.6 eV (34.9%), C-O/C-N at 285.8 eV (35.2%), and a contributed peak derived from epoxy groups at 286.7 eV due to the initial oxidation treatment of the PCC representing 15.5% of the carbon spectra. The amide groups due to the covalent interaction of Tmbgl and the carboxyl groups of the PCC are centered to 288.2 eV (8.4%), while some unaffected carboxyl moieties are observed at 289.4 eV (6.1%). Lastly, for PCCox-Tmbgl-nc (Figure 6d), fitted peaks from C=C/C-H at 284.6 eV (23.8%), C-O/C-N at 285.8 eV (35.5%), and epoxy groups at 287.0 eV (24.7%) are observed. There is no contributed peak from amide bonds, which reveals the non-covalent nature of the interaction between the enzyme and the PCC nanostructure. A peak at 289.0 eV is due to carboxyl groups and represents 11.6% of the whole carbon contribution. Compared to PCCox-Tmbgl-cov, the contribution from carboxyl groups is higher because the enzyme does not interact covalently with them. A last fitted peak at 290.3 eV may arise from π-π interactions between the enzyme and PCC domains or enzyme-enzyme interactions [60]. The atomic percentages are presented in Table 2, while the C/N ratios of covalently or non-covalently immobilized Tmbgl on PCC and PCCox are displayed in Table 3.

Table 2. Atomic percentages of PCC-Tmbgl-cov, PCC-Tmbgl-nc, PCCox-Tmbgl-cov, and PCCox-Tmbgl-nc.

Sample	Atomic Percentage %		
	C	O	N
PCC-Tmbgl-cov	75.1 ± 3.0	14.1 ± 1.1	10.8 ± 0.9
PCC-Tmbgl-nc	75.1 ± 3.0	13.5 ± 1.1	11.4 ± 0.9
PCCox-Tmbgl-cov	55.5 ± 2.2	40.7 ± 2.4	3.8 ± 0.3
PCCox-Tmbgl-nc	72 ± 2.9	23 ± 1.4	5 ± 0.4

Table 3. The C/N ratio of PCC-Tmbgl-cov, PCC-Tmbgl-nc, PCCox-Tmbgl-cov, and PCCox-Tmbgl-nc.

Sample	C/N Ratio
PCC-Tmbgl-cov	6.9
PCC-Tmbgl-nc	6.6
PCCox-Tmbgl-cov	14.6
PCCox-Tmbgl-nc	14.4

The immobilization of β-glucosidase on pristine and oxidized PCCs was confirmed via AFM (Figure 7). Representative AFM images indicate that molecules of β-glucosidase are attached on the surface of the PCC and PCCox. As revealed from the cross-section analysis, the size of the PCC is significantly increased, which verifies the successful attachment of the enzyme. The average thickness of PCC-Tmbgl-cov is 20 to 25 nm. In addition, the average thickness of PCCox-Tmbgl-cov is slightly decreased and has an average value of 18 to 23 nm. From these images, it is clear that the oxidation procedure affects the average size of the cuboids' nanoparticles and also the estimated size of the nano-biocatalytic system.

3.4. Thermal Stability of Free and Immobilized β-Glucosidase

The thermal stability of free and immobilized β-glucosidases was investigated. The half-life time (the time required for the enzyme to lose 50% of its initial activity) was determined after incubation at 60 °C, in citrate phosphate buffer 100 mM, pH 5.0, and pH 6.5 for Albgl and Tmbgl, respectively. Table 4 demonstrates that the half-life time of covalently immobilized Albgl is enhanced by about 50%. These results indicate that the immobilized Albgl exhibits an increased resistance toward thermal denaturation that might be induced by heating. Similar enhanced thermal stability of immobilized β-glucosidase on various materials has been reported [35,61,62]. In the case of Tmbgl, both enzyme forms (free or immobilized) exhibited very high stability at 60 °C, as expected due to the well-known thermal stability of this enzyme. For instance, covalently immobilized Tmbgl on PCC or PCCox preserves more than 95% of its initial activity after incubation for 24 h at 60 °C.

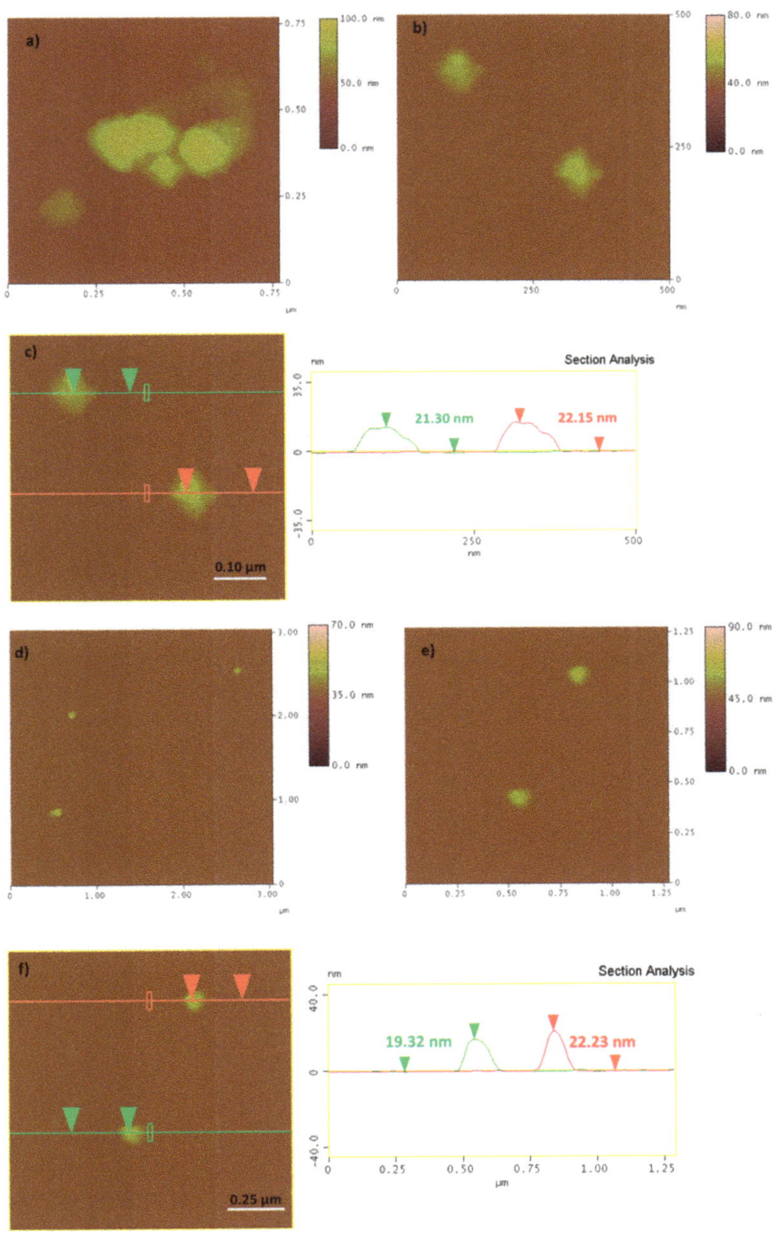

Figure 7. AFM height images and cross section analysis of PCC-Tmbgl (**a–c** images) and PCCox -Tmbgl (**d–f** images).

Table 4. Half-life time of free and immobilized Albgl on PCC and PCCox after incubation at 60 °C in citrate phosphate buffer 100 mM pH 5.0.

Sample	Half-Life Time (Hours)
Albgl	0.6 ± 0.04
PCC-Albgl-cov	0.9 ± 0.10
PCC-Albgl-nc	0.7 ± 0.07
PCCox-Albgl-cov	0.9 ± 0.06
PCCox-Albgl-nc	0.8 ± 0.05

3.5. *Use of Immobilized β-Glucosidase for the Conversion of Oleuropein to Hydroxytyrosol*

Herein, free and immobilized β-glucosidases on PCCox were further used for the chemoenzymatic transformation of OLE to HT. More specific, as described in Figure 1, β-glucosidase catalyzes the hydrolysis of OLE to OLE aglycon (step 1), which, in aqueous solution at a pH of 7.0 and at 60 °C, it further undergoes a fast chemical re-arrangement, which leads to the formation of HT [23,63]. HPLC was used for the quantitative analysis of the reactants and products that were formed in the reaction mixture. The formation of the HT was also followed through liquid chromatography–mass spectrometry (LC-MS) analysis (Figures S1 and S2, Table S1, Supplementary Materials). Figure 8a illustrates the reaction progress of the enzymatic hydrolysis of OLE catalyzed by free and immobilized β-glucosidase. As can be seen in Table 5, free β-glucosidases demonstrate lower conversion rates than immobilized β-glucosidases, which could be attributed to the higher stability of the immobilized β-glucosidases compared to the free enzyme, in a similar manner as previously described (Table 4). It is worth noting that PCCox-Tmbgl preparations present higher hydrolytic activity than that observed for PCCox-Albgl, which is in accordance with that described in Table 5 for the hydrolysis of *p*-NPG. For all the tested bio-nanoconjugates, the OLE hydrolysis yield after 24 h of incubation exceeds 90%, which is higher than that reported for other β-glucosidase preparations under similar reaction conditions [27]. The progress of the formation of HT after the hydrolysis of OLE and the incubation of the reaction mixture at 60 °C at pH 7.0, is described in Figure 8b. The chemoenzymatic conversion of OLE that is described in the present work, results in the formation of 2.4 g L^{-1} of HT, which is among the highest reported concentrations in literature [27,29,64].

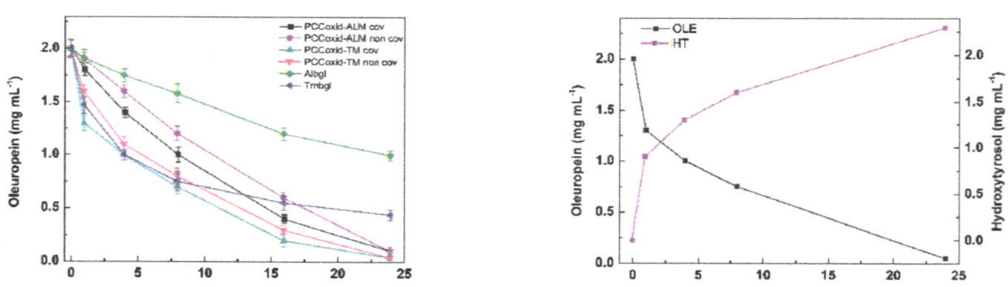

(a) (b)

Figure 8. (**a**) Hydrolysis of OLE catalyzed by immobilized β-glucosidase at different time intervals. The amount of the enzyme in the reaction system was 0.15 mg/mL and 0.07 mg/mL for Albgl and Tmbgl respectively. (**b**) Reaction progress of the enzymatic hydrolysis of OLE catalyzed by PCCox-Tmbgl-cov and the formation of HT at different time intervals (the standard deviation was less than 5% in all cases).

Table 5. Initial reaction rates of immobilized β-glucosidase on PCC and PCCox for the hydrolysis of oleuropein (OLE) and percentage conversion yield of OLE after 24 h of incubation in citrate phosphate buffer 100 mM, pH 5.0, and pH 6.5 for Albgl and Tmbgl, respectively, at 37 °C (the standard deviation was less than 5% in all cases).

Sample	Initial Reaction Rate mM h^{-1} g^{-1} of Biocatalyst	% Conversion Yield of OLE
Free Albgl	0.05	50
PCCox-Albgl-cov	0.18	92
PCCox-Albgl-nc	0.16	90
Free Tmbgl	0.14	78
PCCox-Tmbgl-cov	0.20	98
PCCox-Tmbgl-nc	0.19	95

3.6. *Reusability of Immobilized β-Glucosidase*

Reuse of enzymes is pivotal for large-scale biocatalytic processes, especially from an economical point of view. In this work, we investigated the operational stability of PCCox-Albgl-cov and PCCox-Tmbgl-cov in multiple reaction cycles for the hydrolysis of OLE. Each reaction cycle was performed in citrate phosphate buffer 100 mM pH 5.0 and pH 6.5 for Albgl and Tmbgl, respectively, at 37 °C for 24 h. After each cycle, the biocatalyst was separated through centrifugation, washed with buffer, and reused. Figure 9 illustrates the remaining activity of immobilized β-glucosidases after each reaction cycle. After 10 cycles of reuse (240 h of total operation), the remaining activity of immobilized Albgl is reduced to 20%, which is similar to that observed when other nanomaterials were used as immobilization supports [35,49,62,65]. FTIR spectra analysis was employed in order to investigate possible conformational changes of the immobilized enzyme after the reuse process. The correlation coefficient r of FTIR spectra of PCCox-Albgl-cov before and after the reuse process is 0.94, which indicates that slight conformational changes might occur on the immobilized enzyme, which could explain the loss of its catalytic activity observed after repeated use. On the other hand, the remaining hydrolytic activity of immobilized Tmbgl is higher than 90% after 10 catalytic cycles. This indicates that PCCox-Tmbgl-cov bio-nanoconjugate is very stable and can be efficiently used for the biocatalytic conversion of natural compounds, such as OLE.

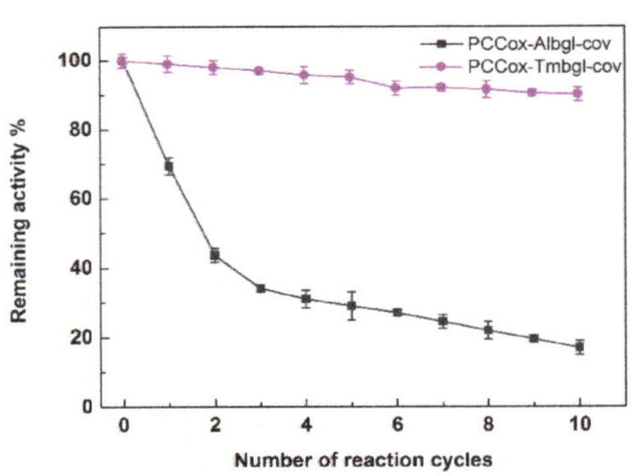

Figure 9. Reusability of covalently immobilized β-glucosidases on PCCox in citrate phosphate buffer 100 mM pH 5.0 and pH 6.5 for Albgl and Tmbgl, respectively, at 37 °C when the enzyme is covalently immobilized on PCCox.

4. Conclusions

PCC and functionalized PCCox were used to develop novel nano-biocatalysts through non-covalent and covalent immobilization of β-glucosidases from two different organisms. The oxidation state of the nanomaterials that were used as immobilization supports and the immobilization procedure seem to affect the immobilization yield and the catalytic activity of the immobilized enzymes. The use of oxidized PCCs as a type of nano-support enhances the catalytic activity of the enzyme, which highlights the importance of the functionalization. The immobilized enzyme retains or exhibits higher thermal stability than the free enzyme, but this enhancement is not dependent on the oxidation state of the PCCs. The bio-nanoconjugates that are formed are able to efficiently catalyze the hydrolysis of OLE, which leads to the formation of a significant amount of HT. Immobilized β-Tmbgl on PCCox demonstrates high operational stability, which indicates—along with the excellent thermal stability- that this nano-biocatalyst is an efficient tool for the bioconversion of OLE and other natural compounds of industrial interest.

Supplementary Materials: The following are available online at http://www.mdpi.com/2079-4991/9/8/1166/s1. Figure S1: HPLC chromatography of (a) OLE, (b) conversion of OLE step 1, and (c) conversion of OLE step 2, at 280 nm, Figure S2: (a) Total ion chromatogram (up) and UV chromatogram (down) at 280 nm of OLE (m/z 539) before the conversion. (b) Total ion chromatogram (up) and UV chromatogram (down) at 280 nm of the reaction mixture of the enzymatic conversion of OLE, step 1. (c) Total ion chromatogram (up) and UV chromatogram (down) at 280 nm of the reaction mixture of the conversion of OLE step 2, Table S1. Peak assignments of reaction mixture of conversion of OLE.

Author Contributions: Conceptualization, K.S., H.S. and D.G.; methodology, A.V.C., E.T., K.-M.L., E.G., M.P., N.C., V.G.K., K.S., H.S. and D.G.; validation, A.V.C., E.T., K.-M.L., E.G., M.P., N.C., V.G.K., and K.S.; investigation, A.V.C., E.T., K.-M.L., E.G., M.P., N.C., V.G.K., K.S., H.S. and D.G.; resources, A.V.C., E.T., K.-M.L., E.G., M.P., N.C., V.G.K., K.S., H.S. and D.G.; data curation, K.S., M.P., D.G. and H.S.; writing—original draft preparation, A.V.C., E.G., M.P. and N.C.; writing—review and editing, K.S., H.S. and D.G.; supervision, K.S., H.S. and D.G..; project administration, H.S. and D.G.; funding acquisition, H.S. and D.G.

Funding: This research has been co-financed by the European Regional Development Fund of the European Union and Greek national funds through the Operational Program Competitiveness, Entrepreneurship and Innovation, under the call RESEARCH—CREATE—INNOVATE (project code: T1EDK-01716). ET was supported by the Hellenic Foundation for Research and Innovation (HFRI) and the General Secretariat for Research and Technology (GSRT), under the HFRI PhD Fellowship grant (GA. No. 1829). N.C. and K.M.L. gratefully acknowledge the IKY Foundation for the financial support: «This research is co-financed by Greece and the European Union (European Social Fund—ESF) through the Operational Programme «Human Resources Development, Education and Lifelong Learning» in the context of the project "Strengthening Human Resources Research Potential via Doctorate Research" (MIS-5000432 and MIS-5002567), implemented by the State Scholarships Foundation (IKY)».

Conflicts of Interest: The authors declare no conflict of interest. The funders had no role in the design of the study; in the collection, analyses, or interpretation of data; in the writing of the manuscript, or in the decision to publish the results.

References

1. Bolivar, J.M.; Eisl, I.; Nidetzky, B. Advanced characterization of immobilized enzymes as heterogeneous biocatalysts. *Catal. Today* **2015**, *259*, 66–80. [CrossRef]
2. Zdarta, J.; Meyer, A.S.; Jesionowski, T.; Pinelo, M. A general overview of support materials for enzyme immobilization: characteristics, properties, practical utility. *Catalysts* **2018**, *8*, 92. [CrossRef]
3. Hong, T.; Liu, W.; Li, M.; Chen, C. Recent advances in the fabrication and application of nanomaterial-based enzymatic microsystems in chemical and biological sciences. *Anal. Chim. Acta* **2019**, *1067*, 31–47. [CrossRef] [PubMed]
4. Pavlidis, I.V.; Vorhaben, T.; Tsoufis, T.; Rudolf, P.; Bornscheuer, U.T.; Gournis, D.; Stamatis, H. Development of effective nanobiocatalytic systems through the immobilization of hydrolases on functionalized carbon-based nanomaterials. *Bioresour. Technol.* **2012**, *115*, 164–171. [CrossRef] [PubMed]
5. Adeel, M.; Bilal, M.; Rasheed, T.; Sharma, A.; Iqbal, H.M.N. Graphene and graphene oxide: Functionalization and nano-bio-catalytic system for enzyme immobilization and biotechnological perspective. *Int. J. Biol. Macromol.* **2018**, *120*, 1430–1440. [CrossRef]
6. Li, Y.; Gao, F.; Wei, W.; Qu, J.; Ma, G.; Zhou, W. Enzymatic pore size of macroporous polystyrene microspheres affects lipase immobilization. *J. Mol. Catal. B Enzym.* **2010**, *66*, 182–189. [CrossRef]
7. Bayne, L.; Ulijn, R.V.; Halling, P.J. Effect of pore size on the performance of immobilised enzymes. *Chem. Soc. Rev.* **2013**, *42*, 9000–9010. [CrossRef]

8. Wan, D.; Tian, L.; Li, X.; Li, B.; Zhang, Q. A versatile strategy for enzyme immobilization: Fabricating lipase/inorganic hybrid nanostructures on macroporous resins with enhanced catalytic properties. *Biochem. Eng. J.* **2018**, *139*, 101–108. [CrossRef]
9. Luangon, B.; Siyasukh, A.; Winayanuwattikun, P.; Tanthapanichakoon, W. Enzymatic flow-through immobilization of *Candida rugosa* lipase on hierarchical micro-/macroporous carbon monolith. *J. Mol. Catal. B Enzym.* **2012**, *75*, 80–85. [CrossRef]
10. Wu, J.; Li, X.; Yan, Y.; Hu, Y.; Zhang, Y.; Tang, Y. Protein adsorption onto nanozeolite: Effect of micropore openings. *J. Colloid Interface Sci.* **2013**, *406*, 130–138. [CrossRef]
11. Pavlidis, I.V.; Tsoufis, T.; Enotiadis, A.; Gournis, D.; Stamatis, H. Functionalized multi-wall carbon nanotubes for lipase immobilization. *Adv. Eng. Mater.* **2010**, *12*, 179–183. [CrossRef]
12. Patila, M.; Pavlidis, I.V.; Diamanti, E.K.; Katapodis, P.; Gournis, D.; Stamatis, H. Enhancement of cytochrome c catalytic behaviour by affecting the heme environment using functionalized carbon-based nanomaterials. *Process Biochem.* **2013**, *48*, 1010–1017. [CrossRef]
13. Jin, L.; Yang, K.; Yao, K.; Zhang, S.; Tao, H.; Lee, S.; Liu, Z.; Peng, R. Functionalized graphene oxide in enzyme engineering: A selective modulator for enzyme activity and thermostability. *ACS Nano* **2012**, *6*, 4864–4875. [CrossRef]
14. Pavlidis, I.V.; Patila, M.; Bornscheuer, U.T.; Gournis, D.; Stamatis, H. Graphene-based nanobiocatalytic systems: Recent advances and future prospects. *Trends Biotechnol.* **2014**, *32*, 312–320. [CrossRef]
15. Hao, G.P.; Mondin, G.; Zheng, Z.; Biemelt, T.; Klosz, S.; Schubel, R.; Eychm ller, A.; Kaskel, S. Unusual ultra-hydrophilic, porous carbon cuboids for atmospheric-water capture. *Angew. Chem. Int. Ed.* **2015**, *54*, 1941–1945. [CrossRef]
16. Karageorgou, D.; Thomou, E.; Vourvou, N.T.; Lyra, K.M.; Chalmpes, N.; Enotiadis, A.; Spyrou, K.; Katapodis, P.; Gournis, D.; Stamatis, H. Antibacterial and algicidal effects of porous carbon cuboid nanoparticles. *ACS Omega* **2019**, *4*, 4991–5001. [CrossRef]
17. Staudenmaier, L. Verfahren zur darstellung der graphitsäure. *Eur. J. Inorg. Chem.* **1898**, *32*, 1481–1487. [CrossRef]
18. Stergiou, D.V.; Diamanti, E.K.; Gournis, D.; Prodromidis, M.I. Comparative study of different types of graphenes as electrocatalysts for ascorbic acid. *Electrochem. Commun.* **2010**, *12*, 1307–1309. [CrossRef]
19. Gengler, R.Y.N.; Veligura, A.; Enotiadis, A.; Diamanti, E.K.; Gournis, D.; Józsa, C.; Van Wees, B.J.; Rudolf, P. Large-yield preparation of high-electronic-quality graphene by a langmuir-schaefer approach. *Small* **2010**, *6*, 35–39. [CrossRef]
20. Liaros, N.; Tucek, J.; Dimos, K.; Bakandritsos, A.; Andrikopoulos, K.S.; Gournis, D.; Zboril, R.; Couris, S. The effect of the degree of oxidation on broadband nonlinear absorption and ferromagnetic ordering in graphene oxide. *Nanoscale* **2016**, *8*, 2908–2917. [CrossRef]
21. Singh, G.; Verma, A.K.; Kumar, V. Catalytic properties, functional attributes and industrial applications of β-glucosidases. *3 Biotech* **2016**, *6*, 1–14. [CrossRef]
22. Ahmed, A.; Nasim, F.; Batool, K.; Bibi, A. Microbial β-glucosidase: Sources, production and applications. *J. Appl. Environ. Microbiol.* **2017**, *5*, 31–46. [CrossRef]
23. Achmon, Y.; Fishman, A. The antioxidant hydroxytyrosol: Biotechnological production challenges and opportunities. *Appl. Microbiol. Biotechnol.* **2014**, *99*, 1119–1130. [CrossRef]
24. Turck, D.; Bresson, J.; Burlingame, B.; Dean, T.; Fairweather-Tait, S.; Heinonen, M.; Hirsch-Ernst, K.I.; Mangelsdorf, I.; McArdle, H.J.; Naska, A.; et al. Safety of hydroxytyrosol as a novel food pursuant to Regulation (EC) No 258/97. *EFSA J.* **2017**, *15*, 4728.
25. Vilaplana-Pérez, C.; Auñón, D.; García-Flores, L.A.; Gil-Izquierdo, A. Hydroxytyrosol and potential uses in cardiovascular diseases, cancer and AIDS. *Front. Nutr.* **2014**, *1*, 18.
26. Zhang, Z.L.; Chen, J.; Xu, Q.; Rao, C.; Qiao, C. Efficient synthesis of hydroxytyrosol from 3,4-dihydroxybenzaldehyde. *Synth. Commun.* **2012**, *42*, 794–798. [CrossRef]
27. Liu, M.; Yong, Q. Efficient bioconversion of oleuropein from olive leaf extract to antioxidant hydroxytyrosol by enzymatic hydrolysis and high-temperature degradation. *Biotechnol. Appl. Biochem.* **2018**, *5*, 680–689. [CrossRef]
28. Yuan, J.; Wang, C.; Ye, J.; Tao, R.; Zhang, Y. Enzymatic hydrolysis of oleuropein from *Olea europea* (Olive) leaf extract and antioxidant activities. *Molecules* **2015**, *20*, 2903–2921. [CrossRef]

29. Khoufi, S.; Hamza, M.; Sayadi, S. Enzymatic hydrolysis of olive wastewater for hydroxytyrosol enrichment. *Bioresour. Technol.* **2011**, *102*, 9050–9058. [CrossRef]
30. Mazzei, R.; Drioli, E.; Giorno, L. Enzyme membrane reactor with heterogenized β-glucosidase to obtain phytotherapic compound: Optimization study. *J. Memb. Sci.* **2012**, *390–391*, 121–129. [CrossRef]
31. Nikolaivits, E.; Termentzi, A.; Skaltsounis, A.L.; Fokialakis, N.; Topakas, E. Enzymatic tailoring of oleuropein from *Olea europaea* leaves and product identification by HRMS/MS spectrometry. *J. Biotechnol.* **2017**, *253*, 48–54. [CrossRef]
32. Sehgal, D.; Vijay, I.K. A method for the high efficiency of water-soluble carbodiimide-mediated amidation. *Anal. Biochem.* **1994**, *218*, 87–91. [CrossRef]
33. Lau, S.C.; Lim, H.N.; Basri, M.; Reza, H.; Masoumi, F.; Ahmad, A. Enhanced biocatalytic esterification with lipase-immobilized chitosan/graphene oxide beads. *PLoS ONE* **2014**, *9*, 1–10. [CrossRef]
34. Bradford, M.M. A rapid and sensitive method for the quantitation of microgram quantities of protein utilizing the principle dye binding. *Anal. Biochem.* **1976**, *72*, 248–254. [CrossRef]
35. Orfanakis, G.; Patila, M.; Catzikonstantinou, A.V.; Lyra, K.; Kouloumpis, A.; Spyrou, K.; Katapodis, P. Hybrid Nanomaterials of magnetic iron nanoparticles and graphene oxide as matrices for the immobilization of β-glucosidase: Synthesis, characterization and biocatalytic properties. *Front. Mater.* **2018**, *5*, 11. [CrossRef]
36. Prestrelski, S.J.; Tedeschi, N.; Arakawa, T.; Carpenter, J.F. Dehydration-induced conformational transitions in proteins and their inhibition by stabilizers. *Biophys. J.* **1993**, *65*, 661–671. [CrossRef]
37. Secundo, F.; Carrea, G.; Molecolare, R.; Bianco, M. Mono- and disaccharides enhance the activity and enantioselectivity of *Burkholderia cepacia* lipase in organic solvent but do not significantly affect its conformation. *Biotechnol. Bioeng.* **2005**, *92*, 438–446. [CrossRef]
38. Secundo, F.; Barletta, G.L.; Dumitriu, E.; Carrea, G.; Molecolare, R.; Bianco, M. Can an inactivating agent increase enzyme activity in organic solvent? Effects of 18-crown-6 on lipase activity, enantioselectivity and conformation. *Biotechnol. Appl. Biochem.* **2007**, *97*, 12–18. [CrossRef]
39. Tzialla, A.A.; Pavlidis, I.V.; Felicissimo, M.P.; Rudolf, P.; Gournis, D.; Stamatis, H. Lipase immobilization on smectite nanoclays: Characterization and application to the epoxidation of α-pinene. *Bioresour. Technol.* **2010**, *101*, 1587–1594. [CrossRef]
40. Kouloumpis, A.; Thomou, E.; Chalmpes, N.; Dimos, K.; Spyrou, K.; Bourlinos, A.B.; Koutselas, I.; Gournis, D.; Rudolf, P. Graphene/carbon dot hybrid thin films prepared by a modified Langmuir-Schaefer method. *ACS Omega* **2017**, *2*, 2090–2099. [CrossRef]
41. Bourlinos, A.B.; Gournis, D.; Petridis, D.; Szabo, T.; Szeri, A.; Dekany, I. Graphite oxide: Chemical reduction to graphite and surface modification with primary aliphatic amines and amino acids. *Langmuir* **2003**, *19*, 6050–6055. [CrossRef]
42. Spyrou, K.; Potsi, G.; Diamanti, E.K.; Ke, X.; Serestatidou, E.; Verginadis, I.I.; Velalopoulou, A.P.; Evangelou, A.M.; Deligiannakis, Y.; Van Tendeloo, G.; et al. Towards novel multifunctional pillared nanostructures: Effective intercalation of adamantylamine in graphene oxide and smectite clays. *Adv. Funct. Mater.* **2014**, *24*, 5841–5850. [CrossRef]
43. Simons, W.W. *Sadtler Research Laboratories The Sadtler Handbook of Infrared Spectra*; Sadtler Research Laboratories: Philadelphia, PA, USA, 1978; ISBN 9780845600344.
44. Bellamy, L.J. *The Infrared Spectra of Complex Molecules: Volume Two Advances in Infrared Group Frequencies*; Springer: Amsterdam, The Netherlands, 1980; ISBN 9789401165228.
45. Enotiadis, A.; Angjeli, K.; Baldino, N.; Nicotera, I.; Gournis, D. Graphene-based nafion nanocomposite membranes: Enhanced proton transport and water retention by novel organo-functionalized graphene oxide nanosheets. *Small* **2012**, *8*, 3338–3349. [CrossRef]
46. Zygouri, P.; Tsoufis, T.; Kouloumpis, A.; Patila, M.; Potsi, G.; Sevastos, A.A.; Sideratou, Z.; Katsaros, F.; Charalambopoulou, G.; Stamatis, H.; et al. Synthesis, characterization and assessment of hydrophilic oxidized carbon nanodiscs in bio-related applications. *RSC Adv.* **2018**, *8*, 122–131. [CrossRef]
47. Patila, M.; Pavlidis, I.V.; Kouloumpis, A.; Dimos, K.; Spyrou, K.; Katapodis, P.; Gournis, D.; Stamatis, H. Graphene oxide derivatives with variable alkyl chain length and terminal functional groups as supports for stabilization of cytochrome c. *Int. J. Biol. Macromol.* **2016**, *84*, 227–235. [CrossRef]
48. Gao, Y.; Kyratzis, I. Covalent immobilization of proteins on carbon nanotubes using the cross-linker 1-ethyl-3-(3-dimethylaminopropyl)carbodiimide—A critical assessment. *Bioconjug. Chem.* **2008**, *19*, 1945–1950. [CrossRef]

49. Xue, Y.M.; Xu, C.Y.; Hou, J.J.; Li, X.Q.; Cao, Z.G. Enhanced soluble expression of a thermostble β glucosidase from *Thermotoga maritima* in *Escherichia coli* and its applicaton in immobilization 1. *Appl. Biochem. Microbiol.* **2015**, *51*, 306–315. [CrossRef]
50. Lee, Y.; Kwon, O.; Yoon, Y.; Ryu, K. Immobilization of horseradish peroxidase on multi-wall carbon nanotubes and its electrochemical properties. *Biotechnol. Lett.* **2006**, *28*, 39–43. [CrossRef]
51. Gómez, J.M.; Romero, M.D.; Fernández, T.M. Immobilization of β-glucosidase on carbon nanotubes. *Catal. Lett.* **2005**, *101*, 275–278.
52. Çelik, A.; Dinçer, A.; Aydemir, T. Characterization of β-glucosidase immobilized on chitosan-multiwalled carbon nanotubes (MWCNTS) and their application on tea extracts for aroma enhancement. *Int. J. Biol. Macromol.* **2016**, *89*, 406–414. [CrossRef]
53. Patila, M.; Kouloumpis, A.; Gournis, D.; Rudolf, P.; Stamatis, H. Laccase-functionalized graphene oxide assemblies as efficient nanobiocatalysts for oxidation reactions. *Sensors (Switzerland)* **2016**, *16*, 287. [CrossRef]
54. Patila, M.; Diamanti, E.K.; Bergouni, D.; Polydera, A.C.; Gournis, D.; Stamatis, H. Preparation and biochemical characterisation of nanoconjugates of functionalized carbon nanotubes and cytochrome c. *Nanomed. Res. J.* **2018**, *3*, 10–18.
55. Zhou, J.; Zhang, J.; David, A.E.; Yang, V.C. Magnetic tumor targeting of β -glucosidase immobilized iron oxide nanoparticles. *Nanotechnology* **2013**, *24*, 1–12. [CrossRef]
56. Chiaradia, V.; Soares, N.S.; Valério, A.; de Oliveira, D.; Araújo, P.H.H.; Sayer, C. Immobilization of *Candida antarctica* Lipase B on magnetic poly(urea-urethane) nanoparticles. *Appl. Biochem. Biotechnol.* **2016**, *180*, 558–575. [CrossRef]
57. Ayoub, F.D.P.; Caseli, L. Biointerfaces controlling the molecular architecture of lactase immobilized in Langmuir-Blodgett films of phospholipids to modulate the enzyme activity. *Colloids Surf. B Biointerfaces* **2017**, *150*, 8–14. [CrossRef]
58. Junior, R.; Caseli, L. Adsorption and enzyme activity of asparaginase at lipid Langmuir and Langmuir-Blodgett films. *Mater. Sci. Eng. C* **2017**, *73*, 579–584. [CrossRef]
59. Serefoglou, E.; Litina, K.; Gournis, D.; Kalogeris, E.; Tzialla, A.A.; Pavlidis, I.V.; Stamatis, H.; Maccallini, E.; Lubomska, M.; Rudolf, P. Smectite clays as solid supports for immobilization of β-glucosidase: Synthesis, characterization and biochemical properties. *Chem. Mater.* **2008**, *20*, 4106–4115. [CrossRef]
60. Spyrou, K.; Calvaresi, M.; Diamanti, E.K.; Tsoufis, T.; Gournis, D.; Rudolf, P.; Zerbetto, F. Graphite oxide and aromatic amines: Size matters. *Adv. Funct. Mater.* **2015**, *25*, 263–269. [CrossRef]
61. Verma, M.L.; Chaudhary, R.; Tsuzuki, T.; Barrow, C.J.; Puri, M. Immobilization of β-glucosidase on a magnetic nanoparticle improves thermostability: Application in cellobiose hydrolysis. *Bioresour. Technol.* **2013**, *135*, 2–6. [CrossRef]
62. Chen, T.; Yang, W.; Guo, Y.; Yuan, R.; Xu, L.; Yan, Y. Enhancing catalytic performance of β-glucosidase via immobilization on metal ions chelated magnetic nanoparticles. *Enzym. Microb. Technol.* **2014**, *63*, 50–57. [CrossRef]
63. Briante, R.; Cara, F.L.; Tonziello, M.P.; Febbraio, F.; Nucci, R. Antioxidant activity of the main bioactive derivatives from oleuropein hydrolysis by hyperthermophilic β-glycosidase. *J. Agric. Food Chem.* **2001**, *49*, 3198–3203. [CrossRef]
64. Hamza, M.; Sayadi, S. Original article high production of *Aspergillus niger* β-glucosidase at pilot-scale and application for hydroxytyrosol release from olive by-product. *Food Sci. Technol.* **2015**, *50*, 1882–1890.
65. Chang, J.; Lee, Y.; Fang, S.; Park, D.; Choi, Y. Hydrolysis of isoflavone glycoside by immobilization of β-glucosidase on a chitosan-carbon in two-phase system. *Int. J. Biol. Macromol.* **2013**, *61*, 465–470. [CrossRef]

nanomaterials

MDPI

Article

One-Step Synthesis Heterostructured g-C_3N_4/TiO_2 Composite for Rapid Degradation of Pollutants in Utilizing Visible Light

Hui Liu [1,†], Zhi-Guang Zhang [1,2,†], Hong-Wei He [3], Xiao-Xiong Wang [1], Jun Zhang [1], Qian-Qian Zhang [1], Yan-Fu Tong [1], Hong-Ling Liu [1], Seeram Ramakrishna [4], Shi-Ying Yan [1,*] and Yun-Ze Long [1,*]

1 Collaborative Innovation Center for Nanomaterials & Devices, College of Physics, Qingdao University, Qingdao 266071, China; lhqddx@163.com (H.L.); zhangzhiguangphysics@126.com (Z.-G.Z.); wangxiaoxiong69@163.com (X.-X.W.); iamjunzhang@163.com (J.Z.); zhangqianABCABC@126.com (Q.-Q.Z.); 17660945629@163.com (Y.-F.T.); 15666428550@163.com (H.-L.L.)
2 College of Science & Information, Qingdao Agricultural University, Qingdao 266109, China
3 Industrial Research Institute of Nonwovens & Technical Textiles, College of Textiles and Clothing, Qingdao University, Qingdao, 266071, China; hhwpost@163.com
4 Center for Nanofibers & Nanotechnology, Faculty of Engineering, National University of Singapore, Singapore; seeram@nus.edu.sg
* Correspondence: ysy5954418@163.com (S.-Y.Y.); yunze.long@163.com or yunze.long@qdu.edu.cn (Y.-Z.L.); Tel.: +86-130 6121 5975 (S.-Y.Y.); +86-139 5329 0681 (Y.-Z.L.)
† These authors contributed equally to this work.

Received: 19 September 2018; Accepted: 12 October 2018; Published: 16 October 2018

Abstract: To meet the urgent need of society for advanced photocatalytic materials, novel visible light driven heterostructured composite was constructed based on graphitic carbon nitride (g-C_3N_4) and fibrous TiO_2. The g-C_3N_4/TiO_2 (CNT) composite was prepared through electrospinning technology and followed calcination process. The state of the g-C_3N_4 and fibrous TiO_2 was tightly coupled. The photocatalytic performance was measured by degrading the Rhodamine B. Compared to commercial TiO_2 (P25®) and electrospun TiO_2 nanofibers, the photocatalytic performance of CNT composite was higher than them. The formation of CNT heterostructures and the enlarged specific surface area enhanced the photocatalytic performance, suppressing the recombination rate of photogenerated carriers while broadening the absorption range of light spectrum. Our studies have demonstrated that heterostructured CNT composite with an appropriate proportion can rational use of visible light and can significantly promote the photogenerated charges transferred at the contact interface between g-C_3N_4 and TiO_2.

Keywords: electrospinning; g-C_3N_4/TiO_2; heterostructures; visible light; photocatalyst

1. Introduction

Dye wastewater was a kind of industrial organic pollution with large chroma, complex composition and difficult to be biochemically treated. In addition, it caused serious harm to the environment and physical health. With the rapid development of modern society, using of economic, environmental friendly, efficient and convenient photocatalytic technology to control the environmental pollution has attracted widespread attention [1,2]. Among wide variety of research materials, semiconductor materials for its economical, stable, and harmless qualities commonly used in photocatalytic technology, including metal oxides, non-metal oxides, nitrides, sulfides, phosphides et al. [3–7]. As a promising candidate photocatalyst, titanium dioxide (TiO_2) due to its chemical stability, non-toxicity, strong oxidation and reduction, controllable morphology and low-cost properties stands out among the various studied transition metal oxide semiconductors over the past decades. It has been proved that anatase phase TiO_2 was considered the most active phase involving in photocatalytic degradation of pollutions [8]. However, some of the inevitable shortcomings inherent in TiO_2 limited the practical application in photocatalytic process. Firstly, as an n-type semiconductor, TiO_2 has a wide band gap (3.2 eV) and absorbs only 4% of the ultraviolet (UV) light in sunlight [9]. Secondly, the high recombination rate of photogenerated carriers in TiO_2 leads to lower quantum efficiency and thus affects the efficiency of photocatalysis [10]. Therefore, it is urgent to find a new strategy for narrowing the band gap of TiO_2, prolonging the lifetime of photogenerated carriers, and improving the photocatalytic performance of the photocatalyst. To date, many effective strategies have been developed to improve the shortcomings of TiO_2 such as precious metal deposition, metal and non-metal ion doping, semiconductor bonding, surface modification, and so on [11–15]. According to the research in recent years, the formation of heterostructure with semiconductor materials is an effective means to avoid the shortcomings inherent in TiO_2. It can be proved that surfacemodification of TiO_2 with a narrow band gap semiconductor could generate a Type II heterostructure. Aguirre et al. have recently employed Cu_2O coated by TiO_2 for improving photocatalytic stability and performance and provided a z-scheme mechanism of charge transfer. Besides, the Cu_2O/TiO_2 nanocomposites were created to avoid the typical limitation problems found in photocatalysis [16]. In addition, various semiconductor materials of narrow band gap combined with TiO_2 have been successfully synthesized, such as CdS, CuS, MoS_2, etc. [17–19]. This strategy can not only extend the light capture range of TiO_2 to the visible light region, but also separate the photogenerated carriers at the contract interface between two different band gap materials through matched energy levels coupling, thereby enhancing the performance of photocatalytic.

Among the studied various narrow band gap semiconductor materials, the g-C_3N_4 with an indirect band gap of 2.7 eV has attracted extensive attention due to its thermal stability, physicochemical stability, excellent photoelectric transmission property, non-toxic and harmless characteristics [20–22]. In addition, semiconductor g-C_3N_4 was easily available and can be obtained by direct thermal decomposition of precursor materials, for instance, urea, melamine, cyanamide and dicyandiamide [23–26]. Nowadays, several techniques have been reported to combine TiO_2 with g-C_3N_4 to form composite. The prepared TiO_2/g-C_3N_4 composite (g-C_3N_4 and TiO_2 with a certain ratio) can not only extend the absorption spectrum range of TiO_2, but also promote the separation efficiency of photogenerated carriers, thus improve the photocatalytic performance [27,28]. For instance, Lu et al. obtained the C-TiO_2/g-C_3N_4 (CNT) composite by the hydrothermal and calcination methods [29]. According to this method, granular TiO_2 was obtained, which has a smaller specific surface area, resulting in a smaller contact surface with lamellar g-C_3N_4. Han et al. fabricated the g-C_3N_4/Titanium(IV) n-butoxide (TNBT)/PVP nanofibers in using the facile electrospinning technology with the addition proportion precursors for g-C_3N_4 and TiO_2 was 1:15, and then the fibers were calcination to obtain the g-C_3N_4 NSs hybridized N-doped TiO_2 nanofibers (GCN/NT NFs) composite [30]. Although the heterostructured CNT composite can be directly obtained after calcination, the doped g-C_3N_4 content accounted for a limited proportion of the composite material, and the heterostructures formed by the contact between g-C_3N_4 and TiO_2 had few sites, making it difficult to harvest a better photocatalytic effect. Lu et al. reported a mean for combining the layered g-C_3N_4 with the rod-like TiO_2 by using the hydrothermal method and calcination process. In comparison with TiO_2 and g-C_3N_4, the heterostructured TiO_2-based nanorods/g-C_3N_4 (TNRs/g-C_3N_4) performed excellent photocatalytic performance and remarkable optoelectronic characteristics for removing heavy metals and degrading rhodamineB (RhB) in wastewater [31]. However, the means of preparing materials was complicated, which required secondary hydrothermal process. Therefore, a novel one-step synthesis method is needed to prepare the CNT composite with large specific surface area and can combine a large amount of g-C_3N_4 to provide more contact sites points.

In the current work, we developed the novel method to prepare CNT composites based on the flexible electrospinning technology [32–35] and calcination process. TBOT/PVP nanofibers were fabricated by electrospinning method, followed by one-step synthesis a type II heterostructure based on n-type g-C_3N_4 and n-type TiO_2 nanofibers in calcination process. The synthesized porous TiO_2 nanofibers were tightly wrapped on the surface of the g-C_3N_4, and the fibrous TiO_2 with large specific surface area were contacted with the substrate of the lamellar state that was provided by a large amount of g-C_3N_4 to form heterostructure. The heterostructured CNT composite was found available in absorbing the visible light and promoting the separation in photogenerated carriers. Consequently, the CNT composite was promising to be applied in practical environmental protection for removing the organic pollutants.

2. Experimental Section

2.1. Materials

Polyvinyl Pyrrolidone (PVP, $M_W \approx 1300000$) was purchased from Shanghai Aladdin Bio-Chem Technology Co., Ltd. (Shanghai, China). Tetrabutyl orthotitanate (TBOT, CP, 98.0%), tert-butyl alcohol (TBA, CP, 98.0%), melamine (CP, 98.0%), acetic acid (AR, 99.5%), ethanol (CP, 95.0%), silver nitrate ($AgNO_3$, AR, 99.8%) and disodium ethylenediaminetetraacetate (Na-EDTA, AR, 99%) were all bought from Sinopharm Chemical Reagent Limited Company (Beijing, China). Degussa P25 (80% anatase and 20% rutile) was purchased from Evonik Degussa Company (Shanghai, China). All chemical reagents we used as received without further purification.

2.2. Preparation of TBOT/PVP Nanofibers Membrane

0.3 g of PVP, 5 g of alcohol and 2 g of acetic acid were weighed into a prepared 50 mL Erlenmeyer flask and mixed well. Then 3 g TBOT was added into the transparent solution and continued for stirring at 30 min until uniform. Later, 5 mL solution was taken into a syringe and placed on an electrospinning device with a voltage of 10 kV. The metal needle was connected to the anode of the high voltage power supply, and the negative pole of the high voltage power supply was connected to the collector. Followed, the prepared TBOT/PVP nanofibers membrane was collected and dried at 60 °C for 12 h.

2.3. Preparation of g-C_3N_4.

g-C_3N_4 was prepared following the method reported in the literature [36]. Ten g of melamine was uniformly dispersed in an alumina crucible and placed into a muffle furnace by annealing at 550 °C for 4 h to obtain the yellow powder.

2.4. Fabrication of CNT Composites

The as-prepared white TBOT/PVP nanofibers membrane and g-C_3N_4 powder were weighed in various certain ratios and grounded in a mortar for 1 h to acquire the uniform g-C_3N_4/TBOT/PVP composites, as listed in Table 1. Subsequently, the mixture was placed into a crucible and then transferred to a muffle furnace by annealing at 600 °C for 2 h. Anneal operation was to remove the PVP present in the nanofibers for leaving bare TiO_2, meanwhile allowed the fibrous TiO_2 tightly combined with g-C_3N_4. In contrast, the TBOT/PVP nanofibers membrane was also annealed in the muffle furnace under the same conditions to obtain bare TiO_2 nanofibers.

Table 1. Description of various as-prepared g-C_3N_4/TiO_2 (CNT) composite.

Samples	CNT1	CNT2	CNT3	CNT4	CNT5	CNT6	CNT7
g-C_3N_4 (g)	0.25	0.3	0.5	1	2	3	4
TBOT/PVP (g)	1	1	1	1	1	1	1

2.5. Characterization

X-ray diffraction (XRD) patterns were recorded by a Rigaku SmartLab X-ray diffractometer (Rigaku, Tokyo, Japan) using Cu-Kα radiation (λ = 1.54178 Å) with an accelerating voltage at 40 kV, a sweep step was 5° in the 2θ range from 10° to 80°. The morphology and microstructure images of the as-synthesized samples were monitored by a JEOL JSM-7800F field emission scanning electron microscope (SEM) (JEOL, Tokyo, Japan). A JEOL JEM-2100F transmission electron microscopy (TEM) (JEOL, Tokyo, Japan) attached with energy dispersive spectroscopy (EDS) was used for observation the as-prepared samples. X-ray photoelectron spectroscopy (XPS) measurement was taken on a Thermo Scientific Escalab 250Xi system (Thermo Scientific, Shanghai, China) with an Al Kα X-ray source to confirm the surface chemical composition of the samples and the valence state of the contained elements. The Brunauer–Emmett–Teller (BET) specific surface areas of the samples were carried out by Quantachrome Autosorb-IQ-MP/XR nitrogen adsorption apparatus (Quantachrome, Shanghai, China). A Hitachi F-4600 fluorescence spectrometer (Hitachi, Tokyo, Japan) was used to measure the photoluminescence (PL) spectra for studying the recombination efficiency of photogenerated carriers with an excitation wavelength of 320 nm. UV-Vis diffuse reflectance spectra of the as-prepared solid composites photocatalysts (the $BaSO_4$ powder was used as a reflectance standard) and UV-Vis absorbance spectra of the reaction solution were collected on a PERSEE-T9 UV-Vis spectrophotometer (PERSEE, Beijing, China).

2.6. Photocatalytic Performance

The photocatalytic performances of the samples were determined by evaluating the concentration of representative pollutants RhB 5 mg/L in the 50 mL solution after irradiation with an 800 W Xe lamp equipped with the filter at 420 nm (only launched visible light). 50 mg photocatalysts was added to the dye solution and continue to stir for 30 min in dark state before turning on the Xe lamp, ensuring the equilibrium of adsorption desorption is achieved [37,38]. The solution was illuminated with a visible light source went on for 2 h during the entire photocatalytic process. Under the action of flowing cooling water circulation system, after every 15 min, 4 mL of the solution was extracted from quartz tube and centrifuged at 10,000 rpm speed for 10 min to separate the supernatant. The concentration of RhB in solution was detected by the UV-Vis spectrophotometer (Beijing, China) using an absorption wavelength of 540 nm for indicating the photocatalytic performance. In addition, the roles of various active species in the reaction system were also confirmed by the addition of corresponding scavengers.

3. Results and Discussion

3.1. Synthesis and Application Process

The flow chart of the as-prepared photocatalysts from preparation to application was shown in Figure 1. At the beginning, TBOT/PVP nanofibers were obtained by electrospinning technology [32]. And then the yellow g-C_3N_4 was synthesized by calcination the melamine at 550 °C for 4 h, following the method mentioned in the literature [36]. Subsequently, the prepared TBOT/PVP nanofibers were mixed and grinded uniformly with g-C_3N_4 powder in a certain proportion and transferred to a muffle furnace for reannealing. Later, the as-prepared various CNT composites were applied to degrade the contaminant under visible light irradiation.

Figure 1. Schematic illustration for the synthesis and visible light photocatalytic application of the CNT composite.

3.2. Structure and Morphology Characteristics

Figure 2 depicts the XRD patterns of g-C_3N_4, TiO_2 and various CNT composites. As shown in Figure 2 curve a, bare TiO_2 nanofibers exhibited significant diffraction peaks 2θ at 25.4°, 37.0°, 37.9°, 38.7°, 48.3°, 54.1°, 55.4°, 62.9°, 69.1°, 70.3°, and 75.1°, which were consistent with the (101), (103), (004), (112), (200), (105), (211), (204), (116), (220), and (215) crystal faces of anatase TiO_2 (JCPDS 21-1272), respectively [39]. In addition, the crystallized peaks 2θ at 13.1° and 27.5° were related to g-C_3N_4 phase at (001) and (002) crystal faces (JCPDS 87-1526), respectively [40]. As observed in Figure 2 curves b and c, only bare TiO_2 was detected and no other peaks appeared in the XRD patterns, owing to the proportion of added g-C_3N_4 was less. The diffraction peaks of g-C_3N_4 increased markedly with the addition of g-C_3N_4 content (Figure 2 curves d and e), indicating a successful compounded of g-C_3N_4 to TiO_2. This phenomenon was also clearly illustrated by SEM images. In addition, the intensity peaks of TiO_2 become lower for the proportion of g-C_3N_4 increases, proving that the content of g-C_3N_4 could affect the transformation from TBOT to anatase TiO_2. Therefore, the results of XRD can inferred that various CNT composites have been successfully synthesized by the process of electrospinning and calcination.

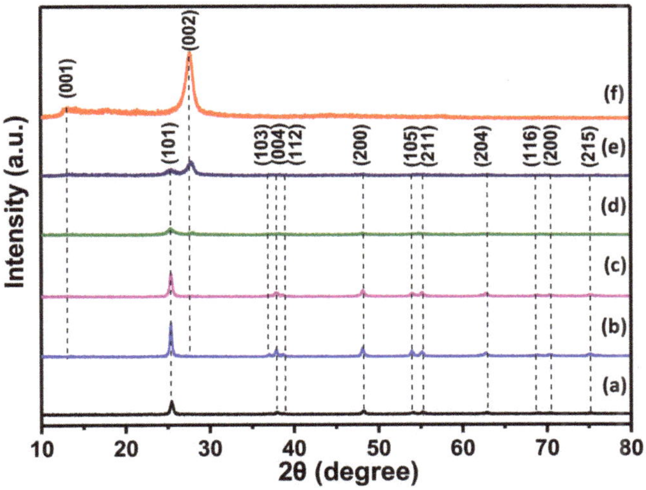

Figure 2. XRD patterns of (**a**) TiO_2 nanofibers; (**b**) CNT1; (**c**) CNT3; (**d**) CNT5; (**e**) CNT6 and (**f**) g-C_3N_4 synthesized at 550 °C.

The detailed structure of heterostructured CNT6 composite has been further confirmed by TEM and HRTEM, as shown in Figure 3a,b, respectively. In TEM image, it could be clearly seen that the fibrous and bulk substances were apparent as TiO_2 and g-C_3N_4, respectively. It seemed that TiO_2 was tightly attached the surface of g-C_3N_4 so the heterostructure formed, although the thickness of the layer of the latter was uncertain. In addition, the TEM image also investigated that the diameter of the nanofiber was less than 100 nm, which was consistent with the results observed in SEM images (Figure S1). In HRTEM image, the well-ordered lattice spacing of TiO_2 was measured at 0.189 nm, which was assigned to the exposed (200) plane of anatase TiO_2 [41]. In addition, the measured lattice spacing of 0.335 nm can be clearly seen, completely ascribed to the (002) plane of g-C_3N_4 [42]. Analysis based on SEM, TEM, HRTEM and XPS results for TiO_2 and g-C_3N_4, the formation of CNT heterostructure between g-C_3N_4 and TiO_2 was verified. The close contact interface may contribute to the transfer and separate of photogenerated carriers during the process of light irradiation, which could enhance the performance of photocatalytic. Therefore, the expectation for improving the photocatalytic performance by CNT composite is reasonable.

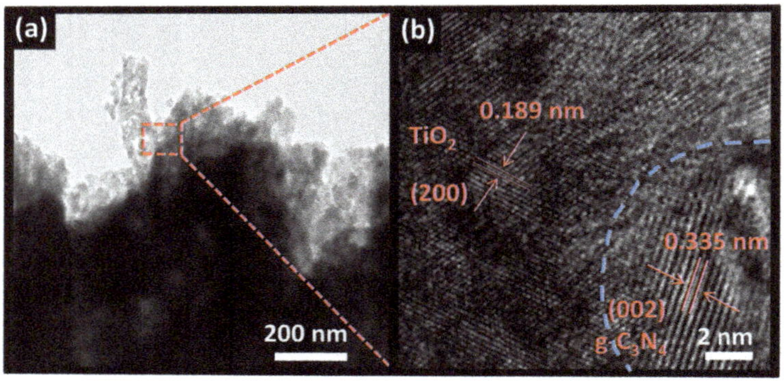

Figure 3. (**a**) TEM and (**b**) HRTEM images of heterostructured CNT6 composite.

The surface chemical composition and elemental states of the as-prepared CNT heterstructured composite were studied using XPS. Figure S2 exhibited the XPS survey spectra included O 1s, Ti 2p, C 1s and N 1s for CNT6 composite. In Figure 4a, O 1s spectrum present two characteristic peaks located at 530.1 eV and 532.2 eV corresponded to Ti–O and C=O, respectively [43]. The spectrum of O 1s indicated that TBOT has been converted to TiO_2 during the sintering process. The C 1s spectrum in Figure 4b exhibited two major peaks centered at 284.8 eV and 288.1 eV belong to C–C and N–C=N, respectively [44]. Regarding of N 1s spectrum (Figure 4c), the basic units of g-C_3N_4 mainly contains three N units, presenting in characteristic peaks to sp^2 hybridized C=N–C (398.7 eV), tertiary nitrogen N–$(C)_3$ (399.6 eV) and C–N–H groups (401.1 eV) [45]. Meanwhile, the peaks of Ti element in Figure 4d distributed at 458.7 eV and 464.5 eV were ascribed to Ti $2p_{3/2}$ and Ti $2p_{1/2}$, respectively [46]. The analysis of XPS confirmed the presence of TiO_2 and g-C_3N_4 in the composites. The results further indicate that the g-C_3N_4/TiO_2 heterostructures have been successfully formed, which was matched well with the results obtained by XRD, TEM, UV and PL.

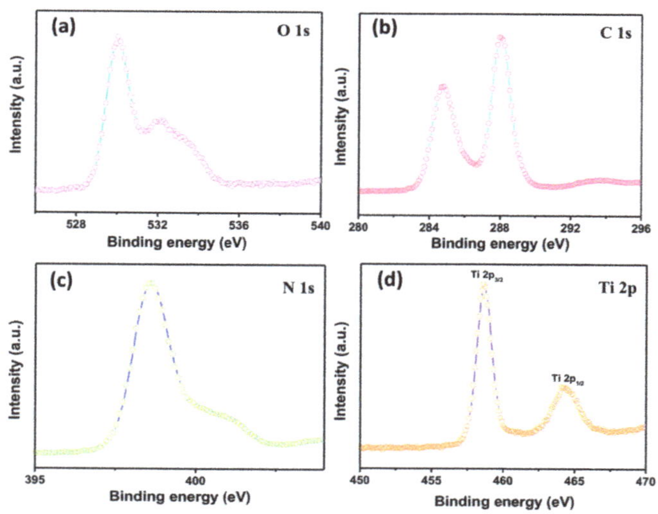

Figure 4. XPS spectra of CNT6: (**a**) O 1s region; (**b**) C 1s region; (**c**) N 1s region and (**d**) Ti 2p region.

Nitrogen adsorption–desorption isotherms analysis was carried out in using BET method to characterize the properties the specific surface area of TiO_2 nanofibers and CNT6 composite

(Figure 5) and the corresponding pore size distribution were recorded by using Barrett–Joyner–Halenda (BJH) method (Figure S3). According to the classification of IUPAC [47], both TiO_2 nanofibers and CNT6 composite adsorption branches curves exhibit the typical type IV isotherm with H3 hysteresis loops in the relative pressure (P/P_0) range about 0.5–1.0, reflecting the slit-like mesopores appear in the products. By BET method, the specific surface area and pore volume of CNT6 were calculated to be 58.71 $m^2\ g^{-1}$ and 0.243 $cm^3\ g^{-1}$, respectively, which was much super than bare TiO_2 nanofibers (34.95 $m^2\ g^{-1}$ and 0.18 $cm^3\ g^{-1}$). The phenomenon might due to the presence of g-C_3N_4, which possessing a large specific surface area and have extensive contract with TiO_2 to inhibit the stack between g-C_3N_4 sheets. It could be deduced that heterostructured CNT6 composite was more suitable for meeting the demand as a photocatalyst than bare TiO_2 nanofibers, owing to the more adsorption reactive contact sites could be provided during the photocatalytic degradation process. Besides, the pore diameter distribution of CNT6 composite (16.53 nm) tends to be smaller compared to TiO_2 nanofibers (20.58 nm), might due to the CNT6 composite surface was covered by TiO_2 nanofibers.

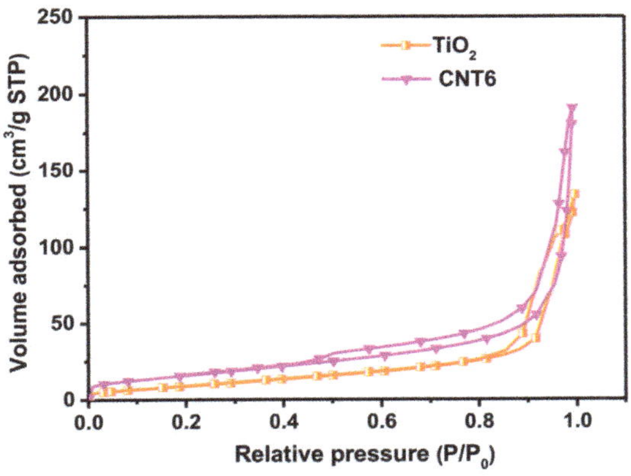

Figure 5. Nitrogen adsorption–desorption isotherms.

3.3. *Optical Characteristics*

The optical absorption properties of g-C_3N_4, TiO_2 and various CNT composites were obtained by UV-Vis diffuse reflectance spectra. It can be seen from Figure 6a that all materials have strong absorption in corresponding response light source region. As expect, g-C_3N_4 with the intrinsic band gap at 2.7 eV exhibits a significant absorption edge at around 450 nm in visible region [48]. The bare TiO_2 nanofibers showed the absorption edge at the wavelength of lower than 400 nm in UV region, ascribing to the band gap of 3.2 eV [49]. After compounding the g-C_3N_4 with TiO_2, the CNT composites present evident hybrid adsorption features that the background to capture visible light obviously increased with the increasing of g-C_3N_4 content in the composites. Obviously, red shift phenomenon has occurred on composites with the increasing of g-C_3N_4 content, performing that the absorption edge moved moderated toward the infrared zone. The situation was also consistent with the color changes of the resulting products from light yellow to black with the increase of g-C_3N_4 added amount as shown in Figure S4. The band gap energy range of various CNT composites was estimated at around 2.3 eV–3.1 eV as displayed in Figure 6b. It was calculated by extrapolating the linear region of $(h\upsilon \cdot F(R))^{1/2}$ versus $h\upsilon$ plot to zero F(R), where $h\upsilon$ is the incident photon energy and values of F(R) can be obtained by using the Kubelka–Munk function: $F(R) = (1-R)^2/2R$ [50,51]. The initial curves of $(h\upsilon \cdot F(R))^{1/2}$ versus $h\upsilon$ were originated from the diffuse reflectance spectra as shown in Figure S5. Thus, it can be concluded that the enhanced ability of CNT composites for capturing the light source could contribute to the improvement of photocatalytic performance.

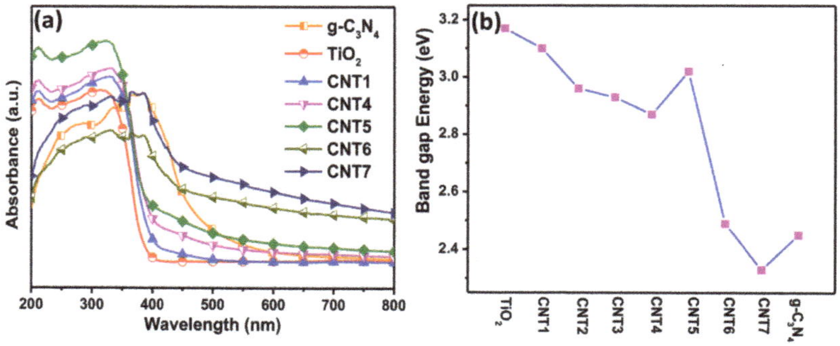

Figure 6. (**a**) UV-vis diffuses reflectance spectra of g-C_3N_4, TiO_2 nanofibers and various CNT composites; (**b**) The curve of band gap energy in various samples.

Since the PL emission is derived from the recombination of photogenerated carriers, PL analysis is an effective technique used in photocatalytic process to investigate the transfer, migrate, and separate of electrons and holes. PL spectra of TiO_2 nanofibers and various CNT composites were recorded under the excitation wavelength of 320 nm (Figure 7). For pure g-C_3N_4, one strong peak centered could see at 473.2 nm was ascribed to the band-band PL phenomenon with the light energy bordering on the band gap energy of g-C_3N_4 [52]. The PL intensities of all CNT composites were lower than pure g-C_3N_4, one reason is that the formation of heterostructures indeed promote the separation of photogenerated carriers at the integrated interface, another reason probably because that g-C_3N_4 occupied is less than the single g-C_3N_4 in the equivalent amount of the products. Comparing the curves of CNT6, CNT7 and g-C_3N_4, it was not difficult to find that the intensity of the emission peak has dropped sharply, which was related to the coupled amount of TiO_2. Further study the enlarged view of partial curves in Figure 7 as depicted in Figure S6, the PL intensity of CNT1, CNT2, CNT3, CNT4, CNT5 were both decreased than bare TiO_2, even the PL intensity of CNT3, CNT4, CNT5 were below than the commercial P25®. Usually, the PL intensity is inversely to the recombination efficiency of the photogenerated carriers. In other words, the lower the PL intensity, the longer the lifetime of the electrons and holes are, and the better performance performs of the corresponding photocatalyst [53]. Therefore, a suitable ratio of CNT composite is beneficial to improve the photocatalytic performance towards degrading pollutants.

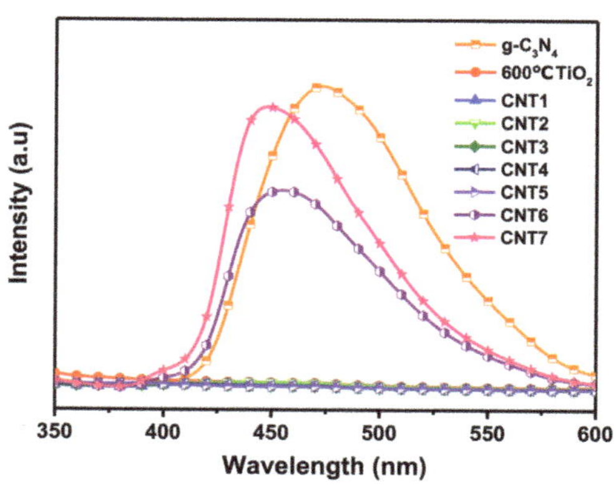

Figure 7. PL spectra of TiO_2 nanofibers and various CNT composites.

3.4. Photocatalytic Performances

The photocatlytic performances of the as-prepared samples were evaluated by monitoring the concentration changes in RhB over time under visible light irradiation. In addition, the system was adsorbed in the dark for 30 min to ensure an adsorption-desorption balance achieved. Figure 8a presents the comparison curves of the concentration in RhB degradation by pure g-C_3N_4, bare TiO_2 nanofibers, commercial P25® and partial as-prepared CNT composites within the same time. After dark state adsorption, CNT6 exhibits the best adsorption effect due to the large specific surface area. Figure 8a showed that the concentration changes of RhB can be neglected throughout the photocatalytic process in the absence of photocatalyst, indicating that RhB was relatively stable under visible light irradiation. It can be seen from Figure 8a that the CNT composites reaction system has better photocatalytic performance for degradation of RhB than single material and commercial P25® reaction systems, which was attributed to the effective separation of photogenerated carriers at the intimate contact interface and the enlarged specific surface area. Surprisingly, the wide band gap of TiO_2 has restricted the range of light absorption wavelength, while it still has good photocatalytic performance in the visible light range. One detailed reason could ascribe that the RhB chromosphere absorbed the visible light and caused the electrons from ground state jumped to the excited state. Subsequently, the excited state electrons rapidly transferred to the conduction band (CB) of TiO_2, resulting in degradation of RhB [54]. Another reason might owing to the material inherent oxygen vacancy defects and thus altered the range of absorbed light to degradation the RhB. As expect, the as-prepared heterostructured CNT6 showed the optimal photocatalytic performance instead of the as-prepared maximum g-C_3N_4 amount loaded in the composite (Figure S7). Hence, it should be noted that the suitable amount of g-C_3N_4 bonded in the composite could greatly affect the performance of photocatalysis. The results indicated that the as-prepared product had excellent photocatalytic performance and was suitable as a photocatalyst for harnessing the actual environmental pollution. The photocatalytic degradation of RhB could be expressed as a pseudo-first-order kinetics process, which followed the equation below: [55]

$$-\ln(C/C_0) = k_{app}t$$

where k_{app} was the reaction rate constant, and C and C_0 were the concentration of RhB detected at initial t_0 and t, respectively. Figure 8b displayed the reaction rate constant of degradation RhB over different as-prepared photocatalysts, which was stemmed from Figure S7. According to the equation, the reaction rate constant of CNT6 was calculated as 9.83×10^{-3} min^{-1}, which was nearly 2.5 times and 3 times higher than that of TiO_2 nanofibers (4.21×10^{-3} min^{-1}) and commercial P25® (3.41×10^{-3} min^{-1}), respectively.

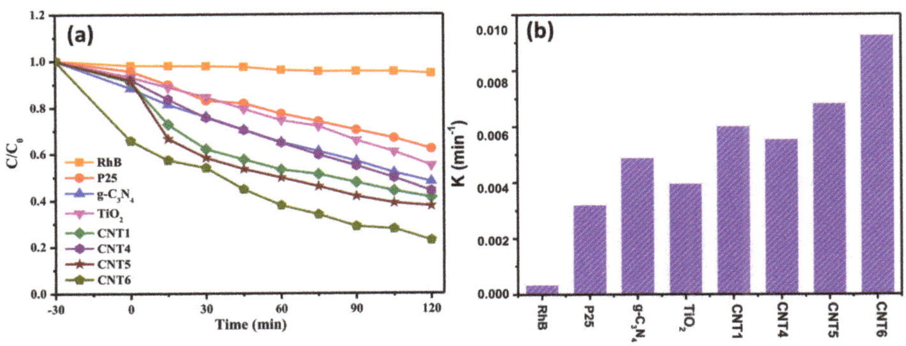

Figure 8. (**a**) Photocatalytic degradation RhB curves under visible light irradiation over different samples; (**b**) reaction rate constant for degradation RhB over different as-prepared photocatalysts.

3.5. Reaction Mechanisms

To further study the possible reaction mechanism of the as-prepared photocatalyst, series comparative experiments of radical scavengers were carried out in the reaction system to demonstrate the active species that dominated the photocatalysis process. Added scavengers of N_2, EDTA-2Na, $AgNO_3$ and TBA were correspond for quenching active species of O_2^-, h^+, e^- and OH, respectivity. As can be seen in Figure 9a, the performances of photocatalysis both have the minor changes in the addition of EDTA-2Na, $AgNO_3$ and TBA, respectively, meaning that h^+, e^- and OH played a weaker role in the decomposition of RhB. When N_2 was added into the solution, the remove rate of RhB over CNT6 reduced drastically, suggesting that O_2^- played the leading role in the photocatalytic process.

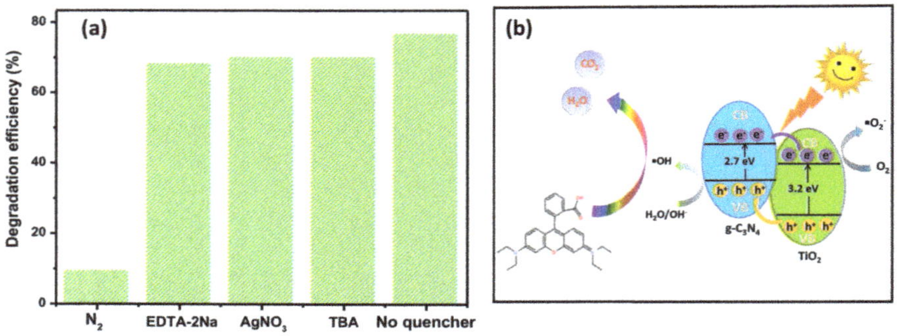

Figure 9. (**a**) Effects on photocatalytic performance by adding different kinds of scavengers; (**b**) Schematic illustration for the separation and recombination process of photogenerated carriers between TiO_2 and g-C_3N_4 under visible light irradiation.

Based on the above results, a reasonable photocatalytic mechanism of heterostructured CNT for degradation RhB under visible light irradiation was proposed in detail, as displayed in Figure 9b. Under the irradiation of visible light, the photogenerated electrons (e^-) of g-C_3N_4 and TiO_2 were both excited from valence band (VB) to CB, leaving the same amount of holes (h^+) on the VB. Since the conduction edge potential of TiO_2 was higher than that of g-C_3N_4, the electrons of g-C_3N_4 on CB were then quickly flowed into the CB of TiO_2 through the contact interface. Correspondingly, the holes on VB of TiO_2 were rapidly transferred into the VB of g-C_3N_4 because the VB potential of former was more negative than latter. Eventually, photogenerated carriers were separated at the intimate interface, thus prolonging the survival time of electrons and holes. The oxygen radicals (O_2^-) attached to the surface of the photocatalyst were generated from that the electrons have transferred to the CB of TiO_2 been captured by O_2 in the solution, possessing strong oxidizing properties on the surface of the photocatalyst ($O_2 + e^- \rightarrow O_2^-$) [56]. Accordingly, the holes moved to the VB of g-C_3N_4 having strong oxidative could react with H_2O or O_2 to yield hydroxyl radicals $(OH)(H_2O, OH^- + h^+ \rightarrow OH)$ [57]. Oxygen radicals and hydroxyl radicals possessed strong oxidizing, can effectively oxidized the organic pollutant RhB in solution transferred into H_2O and CO_2. The mechanism of degradation pathway for RhB was proposed. The degradation of RhB mainly via two steps: N-demethylation and cracking of the conjugated structure. After the main steps of chromophore crack, ring opening and mineralization were completed in the photocatalytic process, the dye was converted to smaller organic species. Finally, the products were mineralization with other organic functional groups to form H_2O and CO_2 [58,59]. It is worth noting that the content of g-C_3N_4 in CNT composites has a significant influence on the separation and recombination efficiency of photogenerated carriers, which has been discussed in PL spectra. An optimal ratio of CNT composite could behave the outstanding photocatalytic performance towards degradation of pollutants under the visible light irradiation.

4. Conclusions

In this paper, novel heterostructured g-C_3N_4/TiO_2 (CNT) composites were successfully fabricated by using g-C_3N_4 precursor and TBOT/PVP nanofibers via the electrospinning technology [60–64], grinding treatment followed by the calcination process. Under the visible light irradiation, CNT composites exhibited considerable photocatalytic performance on the photocatalytic process for purifying organic contaminant as comparing with commercial TiO_2 P25® and pure electrospun TiO_2 nanofibers. An appropriate amount of g-C_3N_4 addition can expand the absorption light range of the composites to the visible light region and enlarge the specific surface area of as-prepared materials. The enhanced performance of the photocatalyst was attributed to the formation of heterostructures between g-C_3N_4 and TiO_2 nanofibers, thus the charges on tightly-coupled interface can quickly transferred and separated. Based on the admirable photocatalytic performance of CNT composite in the process of photocatalysis, the as-prepared products can be treated as prospective materials to government the pollutants in environmental.

Supplementary Materials: The following are available online at http://www.mdpi.com/2079-4991/8/10/842/s1, Figure S1: SEM images of (a) g-C_3N_4, (b) TiO_2 nanofibers, (c,d) CNT6 composite, Figure S2: XPS survey spectra of CNT6, Figure S3: The pore size distribution of TiO_2 nanofibers and CNT6 composite, Figure S4: Optical images of different products: (a) bare TiO_2 nanofibers, (b) g-C_3N_4, (c) CNT2, (d) CNT4, (e) CNT6, and (f) CNT7, Figure S5: The curves of $(h\nu \cdot F(R))^{1/2}$ versus hν originated from the diffuse reflectance spectra, Figure S6: PL spectra of commercial TiO_2 P25 and the enlarged view of partial curves in Figure 6, Figure S7: (a) Photocatalytic degradation RhB curves under visible light irradiation over different samples, (b) kinetic curves of degradation RhB over different as-prepared photocatalysts.

Author Contributions: Conceptualization, H.L. and Z.-G.Z.; Software, H.L.; Validation, H.-W.H. and H.L.; Formal Analysis, Z.-G.Z. and X.-X.W.; Investigation, Q.-Q.Z. and Y.-F.T.; Resources, S.R., S.-Y.Y. and Y.-Z.L.; Data Curation, H.L. and H.-L.L; Writing-Original Draft Preparation, H.L.; Writing-Review & Editing, H.L., Z.-G.Z. and Y.-Z.L.; Supervision, S.R., J.Z. and Y.-Z.L.

Funding: Our work was supported by the National Natural Science Foundation of China (51673103), Shandong Provincial Natural Science Foundation, China (ZR2017BA013), China Postdoctoral Science Foundation (2017M612200) and the Postdoctoral Scientific Research Foundation of Qingdao (2016014).

Acknowledgments: All results were obtained at the Qingdao University. We thank two anonymous reviewers for positive and instructive comments.

Conflicts of Interest: The authors declare no conflict of interest.

References

1. Ding, Q.; Meng, F.; English, C.R.; Cabán-Acevedo, M.; Shearer, M.J.; Liang, D.; Daniel, A.S.; Hamers, R.J.; Jin, S. Efficient photoelectrochemical hydrogen generation using heterostructures of Si and chemically exfoliated metallic MoS_2. *J. Am. Chem. Soc.* **2014**, *136*, 8504–8507. [CrossRef] [PubMed]
2. Zou, Z.; Ye, J.; Sayama, K.; Arakawa, H. Direct splitting of water under visible light irradiation with an oxide semiconductor photocatalyst. *Nature* **2001**, *414*, 625–627. [CrossRef] [PubMed]
3. De Brito, J.F.; Tavella, F.; Genovese, C.; Ampelli, C.; Zanoni, M.V.B.; Centi, G.; Perathoner, S. Role of CuO in the modification of the photocatalytic water splitting behavior of TiO_2 nanotube thin films. *Appl. Catal. B Environ.* **2018**, *224*, 136–145. [CrossRef]
4. Odenbrand, C.U.I. $CaSO_4$ deactivated V_2O_5-WO_3/TiO_2 SCR catalyst for a diesel power plant. Characterization and simulation of the kinetics of the SCR reactions. *Appl. Catal. B Environ.* **2018**, *234*, 365–377. [CrossRef]
5. Hao, R.; Wang, G.; Tang, H.; Sun, L.; Xu, C.; Han, D. Template-free preparation of macro/mesoporous g-C_3N_4/TiO_2 heterojunction photocatalysts with enhanced visible light photocatalytic activity. *Appl. Catal. B Environ.* **2016**, *187*, 47–58. [CrossRef]
6. Qin, N.; Xiong, J.; Liang, R.; Liu, Y.; Zhang, S.; Li, Y.; Li, Z.; Wu, L. Highly efficient photocatalytic H_2 evolution over MoS_2/CdS-TiO_2 nanofibers prepared by an electrospinning mediated photodeposition method. *Appl. Catal. B Environ.* **2017**, *202*, 374–380. [CrossRef]

7. Castro, Y.; Mosa, J.; Aparicio, M.; Pérez-Carrillo, L.A.; Vílchez, S.; Esquena, J.; Durán, A. Sol–gel hybrid membranes loaded with meso/macroporous SiO_2, TiO_2–P_2O_5 and SiO_2–TiO_2–P_2O_5 materials with high proton conductivity. *Mater. Chem. Phys.* **2015**, *149–150*, 686–694. [CrossRef]
8. Hirano, M.; Nakahara, C.; Ota, K.; Tanaike, O.; Inagaki, M. Photoactivity and phase stability of ZrO_2-doped anatase-type TiO_2 directly formed as nanometer-sized particles by hydrolysis under hydrothermal conditions. *J. Solid State Chem.* **2003**, *170*, 39–47. [CrossRef]
9. Ren, W.; Ai, Z.; Jia, F.; Zhang, L.; Fan, X.; Zou, Z. Low temperature preparation and visible light photocatalytic activity of mesoporous carbon-doped crystalline TiO_2. *Appl. Catal. B Environ.* **2007**, *69*, 138–144. [CrossRef]
10. Takezawa, Y.; Imai, H. Bottom–Up synthesis of Titanate nanosheets with hierarchical structures and a high specific surface area. *Small* **2006**, *2*, 390–393. [CrossRef] [PubMed]
11. Yu, H.; Liu, W.; Wang, X.; Wang, F. Promoting the interfacial H_2-evolution reaction of metallic Ag by Ag_2S cocatalyst: A case study of TiO_2/Ag-Ag_2S photocatalyst. *Appl. Catal. B Environ.* **2018**, *225*, 415–423. [CrossRef]
12. Chen, J.; Qiu, F.; Zhang, Y.; Liang, J.; Zhu, H.; Cao, S. Enhanced supercapacitor performances using C-doped porous TiO_2 electrodes. *Appl. Surf. Sci.* **2015**, *356*, 553–560. [CrossRef]
13. Kondo, K.; Murakami, N.; Ye, C.; Tsubota, T.; Ohno, T. Development of highly efficient sulfur-doped TiO_2 photocatalysts hybridized with graphitic carbon nitride. *Appl. Catal. B Environ.* **2013**, *142–143*, 362–367. [CrossRef]
14. Jiang, Z.; Jiang, D.; Yan, Z.; Liu, D.; Qian, K.; Xie, J. A new visible light active multifunctional ternary composite based on TiO_2–In_2O_3 nanocrystals heterojunction decorated porous graphitic carbon nitride for photocatalytic treatment of hazardous pollutant and H_2 evolution. *Appl. Catal. B Environ.* **2015**, *170–171*, 195–205. [CrossRef]
15. Xu, J.; Wu, F.; Jiang, Q.; Shang, J.K.; Li, Y.X. Metal halides supported on mesoporous carbon nitride as efficient heterogeneous catalysts for the cycloaddition of CO_2. *J. Mol. Catal. A Chem.* **2015**, *403*, 77–83. [CrossRef]
16. Aguirre, M.E.; Zhou, R.; Eugene, A.J.; Guzman, M.I.; Grela, M.A. Cu_2O/TiO_2 heterostructures for CO_2 reduction through a direct Z-scheme: Protecting Cu_2O from photocorrosion. *Appl. Catal. B Environ.* **2017**, *217*, 485–493. [CrossRef]
17. Zazpe, R.; Sopha, H.; Prikryl, J.; Krbal, M.; Mistrik, J.; Dvorak, F.; Hromadko, L.; Macak, J.M. A 1D conical nanotubular TiO_2/CdS heterostructure with superior photon-to-electron conversion. *Nanoscale* **2018**, *2018*. *10*, 16601–16612. [CrossRef]
18. Khanchandani, S.; Kumar, S.; Ganguli, A.K. Comparative study of TiO_2/CuS core/shell and composite nanostructures for efficient visible light photocatalysis. *ACS Sustain. Chem. Eng.* **2016**, *4*, 1487–1499. [CrossRef]
19. Li, P.; Hu, H.; Xu, J.; Jing, H.; Peng, H.; Lu, J.; Wu, C.; Ai, S. New insights into the photo-enhanced electrocatalytic reduction of carbon dioxide on MoS_2-rods/TiO_2 NTs with unmatched energy band. *Appl. Catal. B Environ.* **2014**, *147*, 912–919. [CrossRef]
20. Naseri, A.; Samadi, M.; Pourjavadi, A.; Moshfegh, A.Z.; Ramakrishna, S. Graphitic carbon nitride (g-C_3N_4)-based photocatalysts for solar hydrogen generation: Recent advances and future development directions. *J. Mater. Chem. A* **2017**, *5*, 23406–23433. [CrossRef]
21. Cao, S.; Low, J.; Yu, J.; Jaroniec, M. Polymeric photocatalysts based on graphitic carbon nitride. *Adv. Mater.* **2015**, *27*, 2150–2176. [CrossRef] [PubMed]
22. Ong, W.J.; Tan, L.L.; Ng, Y.H.; Yong, S.T.; Chai, S.P. Graphitic carbon nitride (g-C_3N_4)-based photocatalysts for artificial photosynthesis and environmental remediation: Are we a step closer to achieving sustainability? *Chem. Rev.* **2016**, *116*, 7159–7329. [CrossRef] [PubMed]
23. Wan, Z.; Zhang, G.; Wu, X.; Yin, S. Novel visible-light-driven Z-scheme $Bi_{12}GeO_{20}$/g-C_3N_4 photocatalyst: Oxygen-induced pathway of organic pollutants degradation and proton assisted electron transfer mechanism of Cr(VI) reduction. *Appl. Catal. B Environ.* **2017**, *207*, 17–26. [CrossRef]
24. Bellardita, M.; García-López, E.I.; Marcì, G.; Krivtsov, I.; García, J.R.; Palmisano, L. Selective photocatalytic oxidation of aromatic alcohols in water by using P-doped g-C_3N_4. *Appl. Catal. B Environ.* **2018**, *220*, 222–233. [CrossRef]
25. Da Silva, G.T.S.T.; Carvalho, K.T.G.; Lopes, O.F.; Ribeiro, C. g-C_3N_4/Nb_2O_5 heterostructures tailored by sonochemical synthesis: Enhanced photocatalytic performance in oxidation of emerging pollutants driven by visible radiation. *Appl. Catal. B Environ.* **2017**, *216*, 70–79. [CrossRef]

26. Wang, F.; Wang, Y.; Feng, Y.; Zeng, Y.; Xie, Z.; Zhang, Q.; Su, Y.; Chen, P.; Liu, Y.; Yao, K.; et al. Novel ternary photocatalyst of single atom-dispersed silver and carbon quantum dots co-loaded with ultrathin g-C_3N_4 for broad spectrum photocatalytic degradation of naproxen. *Appl. Catal. B Environ.* **2018**, *221*, 510–520. [CrossRef]
27. Li, K.; Huang, Z.; Zeng, X.; Huang, B.; Gao, S.; Lu, J. Synergetic effect of Ti^{3+} and oxygen doping on enhancing photoelectrochemical and photocatalytic properties of TiO_2/g-C_3N_4 heterojunctions. *ACS Appl. Mater. Inter.* **2017**, *9*, 11577–11586. [CrossRef] [PubMed]
28. Lu, L.; Wang, G.; Zou, M.; Wang, J.; Li, J. Effects of calcining temperature on formation of hierarchical TiO_2/g-C_3N_4 hybrids as an effective Z-scheme heterojunction photocatalyst. *Appl. Surf. Sci.* **2018**, *441*, 1012–1023. [CrossRef]
29. Lu, Z.; Zeng, L.; Song, W.; Qin, Z.; Zeng, D.; Xie, C. In situ synthesis of C-TiO_2/g-C_3N_4 heterojunction nanocomposite as highly visible light active photocatalyst originated from effective interfacial charge transfer. *Appl. Catal. B Environ.* **2017**, *202*, 489–499. [CrossRef]
30. Han, C.; Wang, Y.; Lei, Y.; Wang, B.; Wu, N.; Shi, Q.; Li, Q. In situ synthesis of graphitic-C_3N_4 nanosheet hybridized N-doped TiO_2 nanofibers for efficient photocatalytic H_2 production and degradation. *Nano Res.* **2015**, *8*, 1199–1209. [CrossRef]
31. Lu, D.; Fang, P.; Wu, W.; Ding, J.; Jiang, L.; Zhao, X.; Li, C.; Yang, M.; Li, Y.; Wang, D. Solvothermal-assisted synthesis of self-assembling TiO_2 nanorods on large graphitic carbon nitride sheets with their anti-recombination in the photocatalytic removal of Cr(VI) and rhodamine B under visible light irradiation. *Nanoscale* **2017**, *9*, 3231–3245. [CrossRef] [PubMed]
32. Wang, X.; Song, W.Z.; You, M.H.; Zhang, J.; Yu, M.; Fan, Z.; Ramakrishna, S.; Long, Y.-Z. Bionic single-electrode electronic skin unit based on piezoelectric nanogenerator. *ACS Nano* **2018**, *12*, 8588–8596. [CrossRef] [PubMed]
33. Zhang, J.; Wang, X.X.; Zhang, B.; Ramakrishna, S.; Yu, M.; Ma, J.W.; Long, Y.Z. In situ assembly of well-dispersed Ag nanoparticles throughout electrospun alginate nanofibers for monitoring human breath—Smart fabrics. *ACS Appl. Mater. Inter.* **2018**, *10*, 19863. [CrossRef] [PubMed]
34. Hu, W.; Zhang, B.; Luo, W.; Zhang, J.; Guo, Y.; Chen, S.; Yun, M.; Ramakrishna, S.; Long, Y.Z. Ag/alginate nanofiber membrane for flexible electronic skin. *Nanotechnology* **2017**, *28*, 445502. [CrossRef] [PubMed]
35. Zhang, B.; Zhang, Z.G.; Yan, X.; Wang, X.X.; Zhao, H.; Guo, J.; Feng, J.Y.; Long, Y.Z. Chitosan nanostructures by in situ electrospinning for high-efficiency PM2.5 capture. *Nanoscale* **2017**, *9*, 4154–4161. [CrossRef] [PubMed]
36. Yan, S.C.; Li, Z.S.; Zou, Z.G. Photodegradation performance of g-C_3N_4 fabricated by directly heating melamine. *Langmuir* **2009**, *25*, 10397–10401. [CrossRef] [PubMed]
37. Zhou, R.; Guzman, M.I. CO_2 reduction under periodic illumination of ZnS. *J. Phys. Chem. C* **2014**, *118*, 11649–11656. [CrossRef]
38. Zhou, R.; Guzman, M.I. Photocatalytic reduction of fumarate to succinate on ZnS mineral surfaces. *J. Phys. Chem. C* **2016**, *120*, 7349–7357. [CrossRef]
39. Liu, H.; Zhang, Z.G.; Wang, X.X.; Nie, G.D.; Zhang, J.; Zhang, S.X.; Cao, N.; Yan, S.Y.; Long, Y.Z. Highly flexible Fe_2O_3/TiO_2 composite nanofibers for photocatalysis and utraviolet detection. *J. Phys. Chem. Solids* **2018**, *121*, 236–246. [CrossRef]
40. Sun, Y.; Zhang, W.; Xiong, T.; Zhao, Z.; Dong, F.; Wang, R.; Ho, W.-K. Growth of BiOBr nanosheets on C_3N_4 nanosheets to construct two-dimensional nanojunctions with enhanced photoreactivity for NO removal. *J. Colloid Interf. Sci.* **2014**, *418*, 317–323. [CrossRef] [PubMed]
41. Darabdhara, G.; Boruah, P.K.; Borthakur, P.; Hussain, N.; Das, M.R.; Ahamad, T.; Alshehri, S.M.; Malgras, V.; Wu, K.C.W.; Yamauchi, Y. Reduced graphene oxide nanosheets decorated with Au–Pd bimetallic alloy nanoparticles towards efficient photocatalytic degradation of phenolic compounds in water. *Nanoscale* **2016**, *8*, 8276–8287. [CrossRef] [PubMed]
42. Pan, F.; Cao, Z.; Zhao, Q.; Liang, H.; Zhang, J. Nitrogen-doped porous carbon nanosheets made from biomass as highly active electrocatalyst for oxygen reduction reaction. *J. Power Sources* **2014**, *272*, 8–15. [CrossRef]
43. Liu, C.; Wang, L.; Tang, Y.; Luo, S.; Liu, Y.; Zhang, S.; Zeng, Y.; Xu, Y. Vertical single or few-layer MoS_2 nanosheets rooting into TiO_2 nanofibers for highly efficient photocatalytic hydrogen evolution. *Appl. Catal. B Environ.* **2015**, *164*, 1–9. [CrossRef]

44. Liang, Q.; Li, Z.; Yu, X.; Huang, Z.H.; Kang, F.; Yang, Q.H. Macroscopic 3D porous graphitic carbon nitride monolith for enhanced photocatalytic hydrogen evolution. *Adv. Mater.* **2015**, *27*, 4634–4639. [CrossRef] [PubMed]
45. Jiang, X.H.; Xing, Q.J.; Luo, X.B.; Li, F.; Zou, J.P.; Liu, S.S.; Li, X.; Wang, X.K. Simultaneous photoreduction of Uranium(VI) and photooxidation of Arsenic(III) in aqueous solution over g-C_3N_4/TiO_2 heterostructured catalysts under simulated sunlight irradiation. *Appl. Catal. B Environ.* **2018**, *228*, 29–38. [CrossRef]
46. Zheng, F.; Wang, Z.; Chen, J.; Li, S. Synthesis of carbon quantum dot-surface modified P25 nanocomposites for photocatalytic degradation of p-nitrophenol and acid violet 43. *RSC Adv.* **2014**, *4*, 30605–30609. [CrossRef]
47. Thommes, M.; Kaneko, K.; Neimark, A.V.; Olivier, J.P.; Rodriguezreinoso, F.; Rouquerol, J.; Sing, K.S.W. Physisorption of gases, with special reference to the evaluation of surface area and pore size distribution (IUPAC Technical Report). *Pure Appl. Chem.* **2016**, *38*, 25–25. [CrossRef]
48. Li, G.; Nie, X.; Gao, Y.; An, T. Can environmental pharmaceuticals be photocatalytically degraded and completely mineralized in water using g-C_3N_4/TiO_2 under visible light irradiation?—Implications of persistent toxic intermediates. *Appl. Catal. B Environ.* **2016**, *180*, 726–732. [CrossRef]
49. Chen, Y.; Huang, W.; He, D.; Situ, Y.; Huang, H. Construction of heterostructured g-$C_3N_4/Ag/TiO_2$ microspheres with enhanced photocatalysis performance under visible-light irradiation. *ACS Appl. Mater. Inter.* **2014**, *6*, 14405. [CrossRef] [PubMed]
50. Fujisawa, J.I.; Matsumura, S.; Hanaya, M. A single TiOC linkage induces interfacial charge-transfer transitions between TiO_2 and a π-conjugated molecule. *Chem. Phys. Lett.* **2016**, *657*, 172–176. [CrossRef]
51. Li, R.; Yu, L.; Yan, X.; Tang, Q. Efficient photocatalysts from polymorphic cuprous oxide/zinc oxide microstructures. *RSC Adv.* **2015**, *5*, 11917–11924. [CrossRef]
52. Huang, Z.A.; Sun, Q.; Lv, K.; Zhang, Z.; Li, M.; Li, B. Effect of contact interface between TiO_2 and g-C_3N_4 on the photoreactivity of g-C_3N_4/TiO_2 photocatalyst: (001) vs. (101) facets of TiO_2. *Appl. Catal. B Environ.* **2015**, *164*, 420–427. [CrossRef]
53. Jiang, F.; Yan, T.; Chen, H.; Sun, A.; Xu, C.; Wang, X. A g-C_3N_4–CdS composite catalyst with high visible-light-driven catalytic activity and photostability for methylene blue degradation. *Appl. Surf. Sci.* **2014**, *295*, 164–172. [CrossRef]
54. Pan, L.; Zou, J.J.; Zhang, X.; Wang, L. Water-mediated promotion of dye sensitization of TiO_2 under visible light. *J. Am. Chem. Soc.* **2011**, *133*, 10000–10002. [CrossRef] [PubMed]
55. Tan, C.; Zhu, G.; Hojamberdiev, M.; Okada, K.; Liang, J.; Luo, X.; Liu, P.; Liu, Y. Co_3O_4 nanoparticles-loaded BiOCl nanoplates with the dominant {001} facets: Efficient photodegradation of organic dyes under visible light. *Appl. Catal. B Environ.* **2014**, *152–153*, 425–436. [CrossRef]
56. Chai, B.; Peng, T.; Mao, J.; Li, K.; Zan, L. Graphitic carbon nitride (g-C_3N_4)-Pt-TiO_2 nanocomposite as an efficient photocatalyst for hydrogen production under visible light irradiation. *Phys. Chem. Chem. Phys.* **2012**, *14*, 16745. [CrossRef] [PubMed]
57. He, Y.; Zhang, L.; Teng, B.; Fan, M. New application of Z-Scheme Ag_3PO_4/g-C_3N_4 composite in converting CO_2 to fuel. *Environ. Sci. Technol.* **2015**, *49*, 649–656. [CrossRef] [PubMed]
58. Hisaindee, S.; Meetani, M.A.; Rauf, M.A. Application of LC-MS to the analysis of advanced oxidation process (AOP) degradation of dye products and reaction mechanisms. *TRAC Trend Anal. Chem.* **2013**, *49*, 31–44. [CrossRef]
59. Idrees, F.; Cao, C.; Ahmed, R.; Butt, F.K.; Butt, S.; Tahir, M.; Tanveer, M.; Aslam, I.; Ali, Z. Novel nano-flowers of Nb_2O_5 by template free synthesis and enhanced photocatalytic response under visible light. *Sci. Adv. Mater.* **2015**, *7*, 4204–4220. [CrossRef]
60. Sun, B.; Long, Y.Z.; Zhang, H.D.; Li, M.M.; Duvail, J.L.; Jiang, X.Y.; Yin, H.L. Advances in three-dimensional nanofibrous macrostructures via electrospinning. *Prog. Polym. Sci.* **2014**, *39*, 862–890. [CrossRef]
61. Liu, S.; Liu, S.L.; Long, Y.Z.; Liu, L.Z.; Zhang, H.D.; Zhang, J.C.; Han, W.P.; Liu, Y.C. Fabrication of p-type ZnO nanofibers by electrospinning for field-effect and rectifying devices. *Appl. Phys. Lett.* **2014**, *104*, 042105. [CrossRef]
62. Zhang, B.; Yan, X.; He, H.W.; Yu, M.; Ning, X.; Long, Y.Z. Solvent-free electrospinning: Opportunities and challenges. *Polym. Chem.* **2016**, *8*, 333–352. [CrossRef]

63. Wang, L.; Yang, J.; Ran, B.; Yang, X.; Zheng, W.; Long, Y.Z.; Jiang, X. Small molecular TGF-β1 inhibitor loaded electrospun fibrous scaffolds for preventing hypertrophic scars. *ACS Appl. Mater. Inter.* **2017**, *9*, 32545–32553. [CrossRef] [PubMed]
64. Zhang, Z.G.; Liu, H.; Zhang, B.; Zhang, J.; Liu, R.Z.; Ning, X.; Long, Y.Z. Synthesis and application of highly ordered arrays of TiO_2 rods grown on electrospun PVDF fibers. *Mater. Res. Express* **2017**, *4*, 075907. [CrossRef]

nanomaterials

MDPI

Article

An Emerging Visible-Light Organic–Inorganic Hybrid Perovskite for Photocatalytic Applications

Bianca-Maria Bresolin [1,*], Samia Ben Hammouda [1] and Mika Sillanpää [1,2]

[1] Laboratory of Green Chemistry, School of Engineering Science, Lappeenranta University of Technology, Sammonkatu 12, 50130 Mikkeli, Finland; samiabenhammouda@gmail.com (S.B.H.); Mika.Sillanpaa@lut.fi (M.S.)

[2] Department of Civil and Environmental Engineering, Florida International University, Miami, FL 33174, USA

* Correspondence: biancabresolin@yahoo.it or Bianca-maria.Bresolin@lut.fi

Received: 23 October 2019; Accepted: 28 December 2019; Published: 7 January 2020

Abstract: The development of visible-light active photocatalysts is a current challenge especially energy and environmental-related fields. Herein, methylammonium lead iodide perovskite (MAIPb) was chosen as the novel semiconductor material for its ability of absorbing visible-light. An easily reproducible and efficient method was employed to synthesize the as-mentioned material. The sample was characterized by various techniques and has been used as visible-light photocatalyst for degradation of two model pollutants: rhodamine B (RhB) and methylene-blue (MB). The photo-degradation of RhB was found to achieve about 65% after 180 min of treatment. Moreover, the efficiency was enhanced to 100% by assisting the process with a small amount of H_2O_2. The visible-light activity of the photocatalyst was attributed to its ability to absorb light as well as to enhance separation of photogenerated carriers. The main outcome of the present work is the investigation of a hybrid perovskite as photocatalyst for wastewater treatment.

Keywords: halide perovskite; photocatalysis; visible-light; Rhodamine B; oxidation

1. Introduction

Nowadays, environment pollution and energy related issues captured the attention of new century researchers [1–5]. In particular, accelerated release of pollutants because of a combination of growing population and a rapid industrial development have dramatically increased the water pollution in many parts of the world. On the other hand, it is equally urgent to answer the increasing energy demand and mitigate the negative effect of global warming by means of renewable energy sources. Thus, efficient and eco-friendly methods for the degradation of organic pollutants based on renewable energy source, such as solar light, have become an imperative task worldwide [6–11].

Heterogeneous photocatalysis consists in the dispersion of a solid material, usually a semiconductor, that when irradiated at appropriated wavelengths is capable to generate highly reactive oxygen species (ROS) which can degrade organic pollutants [12,13]. Photocatalysis main advantages are: the room temperature operation, the utilization of clean and renewable solar light as the driving force and any production of hazardous residues after mineralization to align with the "zero" waste scheme for industries [14,15].

Recently, hybrid organic–inorganic halide perovskites (HOIPs) have gain a lot of attention, especially in photovoltaics, because of their remarkable properties. It was 2009 when for the first time Miyasaka and his colleagues employed hybrid perovskites in photovoltaic devices [16]. Then, the studies of the HOIPs have stunned the research community with their remarkable performance and rapid progress [17].

Perovskite general formula is ABX_3. HOIP A-site is occupied by an organic cation, B-site by a metal od group IVA in a divalent oxidation state and X-site by a halogen anion [18]. As reported in previous studies, the electronic properties of the mentioned perovskites is mainly governed by the B-X bonds [19]. Electronic properties are fundamental in the understanding of heterogeneous photocatalysis [20].

Herein, focusing on the compositional, structural, optical, and charges transportation properties, we investigated this class of materials as promising candidate for photocatalytic applications [18,21–25]. First, the advantageous properties are a favorable mobility of the photogenerated charges, a reduced surface recombination and long electron–hole diffusion length because of the strong defect tolerance, the shallow point defects and the benign grain boundary. Second, these materials are known to own an enhanced visible-light shift absorption ability and suitable band gap [23,24]. Moreover, they can be produced by low cost solution processes [26].

According to literature, lead-based HOIPs (MAIPb) has achieved the best efficiency among all the studied hybrid perovksites [27–31]. In MAIPb, A-site is occupied by methyl ammonium cation ($CH_3NH_3^+$), the B-site by lead cation (Pb^{2+}), and the X-site by iodine anion (I^-) (Figure 1) [32,33].

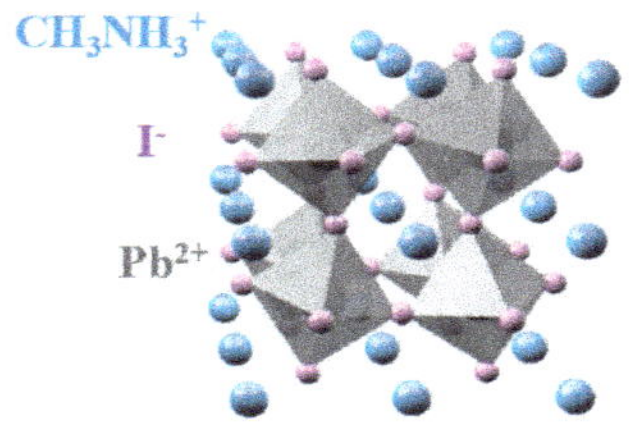

Figure 1. Hybrid organic inorganic perovskite tetragonal structure.

Herein, we propose to determine the feasibility of MAIPb, as one of the most promising HOIPs, as visible-light photocatalyst for the degradation of some dyes having different chemical structures. In particular we investigate the photocatalytic degradation on rhodamine B (RhB, fluorone dye) and methylene blue (MB, thiazine dye) because these dyes are extensively used in industries and medicines [34,35]. Moreover, the effect of key operating conditions on degradation efficiency were studied: catalyst loading, addition of hydrogen peroxide, radiation intensity, solution pH, solution temperature, pollutant initial concentration, and potential recycling test.

2. Experimental

2.1. Materials

Methylamine (CH_3NH_2, 33 wt% in ethanol), hydriodic acid distilled (HI 57 wt% in water), diethyl ether (DE purity ≥ 99.8%), lead (II) iodide (PbI_2 purity 99%), γ-butyrolactone (GBL purity ≥ 99%) were purchased from Sigma Aldrich (Darmstadt, Germany) and used as received. The target dye pollutants RhB, was obtained from Sigma Aldrich (Darmstadt, Germany).

2.2. Photo-Catalyst Synthesis

Hybrid organic-inorganic perovskite was prepared with a one-step, solution-processed method as described in previous literature report [18]. CH_3NH_2 (11.39 mL, 0.09 mol) and HI (10 mL, 0.08 mol) were stirred for 2 h in an ice bath kept at 0 °C to synthesize the precursor, CH_3NH_3I. The solution was evaporated at 50 °C and the solid was washed three times with DE and dried at 50 °C on a hot plate. The CH_3NH_3I (0.39 g) and PbI_2 (1.16 g) were mixed in GBL (10 mL). Finally, the sample was dried at 60 °C for 6 h until the solution was completely evaporated. Before performing photo-catalytic oxidation process, the catalyst was washed several time with deionized water. It should be mentioned that methylammonium was selected as precursor because it is most widely used as A-site cation since its radius appeared to be the more suitable resulting in low packing symmetry and high band gap [36,37]. In comparison to other elements of group IV, Pb was selected because of its performance and stability [18,38–40]. In particular, along group IV from Pb to Ge, it was previously reported a decrease in stability of the divalent oxidation state and a consequent decrease in band gap value combined with a reduced inert electron pair effects [41]. Among the halogens, iodide was selected for its higher efficiency compared to other elements [16]. Moreover, in the periodic table iodide lies close to Pb, thus, they result in more stable structure by sharing similar covalent character [18]. However, we must notice that many factors remain not entirely understood. Moreover, some barriers are still to overcome as stability and toxicity in large-scale implementation.

2.3. Photo-Catalyst Characterization

The X-ray powder diffraction (XRD) spectrum of the catalyst was recorded by PANalytical instrument with the empyrean program (PANalytical, Cambridge, UK) with Co-Kα (λ = 1.7809 Å) as the radiation source, 40 kV generator voltage and 40 mA tube current. The diffraction angle (2θ) ranged from 20°and 80° with intervals of 0.05°. The sample functional groups were characterized by Fourier transform infrared spectra (FT-IR) (Bruker, Solna, Sweden) in the region from 400 to 4000 cm^{-1} at room temperature using Horiba FT-730 FT-IR spectrometer. The microstructure and morphology of the material were defined using scanning electron microscope (SEM) Hitachi SU3500 (Chiyoda, Tokyo, Japan). Energy dispersive spectroscopy (EDS) (Thermo Scientific, Waltham, MA, USA) detected the elemental composition of the pure hybrid organic-inorganic perovskite. The surface composition and the electronic states of elements in the valence-band region were determined by ESCALAB 250 X-ray photoelectron spectroscopy (XPS) (ThermoFisher Scientific, Waltham, MA, USA) with Al-Kα (1486.6 eV) as the X-ray source. Absorption spectra were measured with a PerkinElmer Lambda 1050 spectrophotometer (UV-vis) (PerkinElmer, Waltham, MA, USA) to establish the absorption spectrum and band gap of the sample.

2.4. Procedure for Photo-Catalysis

The visible-light photocatalytic efficiency was evaluated based on the degradation of RhB. All experiments were carried out in Pyrex vessels (100 mL) with 50 mL of RhB (20 mg·L^{-1}). Specified amount of reaction mixture was withdrawn at regular time intervals and analyzed with UV-vis spectrophotometer at emission wavelength of 554 nm [42]. The efficiency of RhB removal was determined as follow:

$$\text{Removal efficiency } \% = C/C_0 \tag{1}$$

where C_0 is the initial concentration of RhB and C is the measured concentration at the time of withdrawal [43,44]. Electron spin resonance (ESR) technique with proper spin traps was used to determine the presence of reactive oxygen species (ROS). TEMP (2,2,6,6-tetramethylpiperidine) was used as spin trap for singlet oxygen and DMSO (dimethyl sulfoxide) for superoxide and hydroxyl radicals [45,46]. The specifics of the visible-light device, used in the current research, are reported in the Supporting Information (Table S1 and Figure S1).

3. Results and Discussion

3.1. Photo-Catalyst Characterization

The morphology of the material was investigated with SEM, the results, shown in Figure 2A,B, suggest an aggregation of nanoparticles with hexagonal shape domains with nanometers size. The specific morphology of the crystal lattice is mainly influenced by the synthesizing temperature and may affect the optical, electrical, and transmission properties of the material, as confirmed in the study of Li et al. [47].

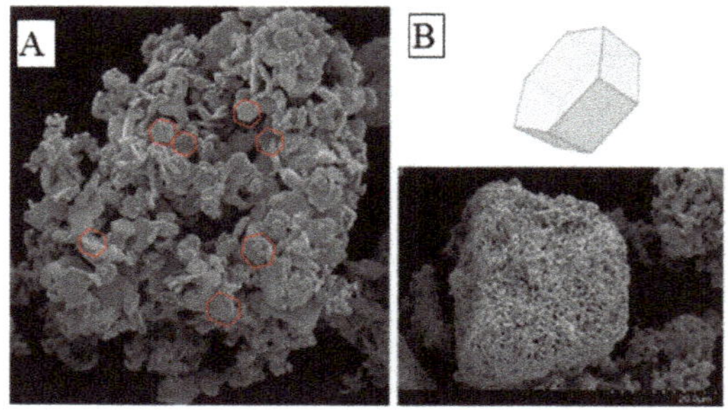

Figure 2. (**A**,**B**) Scanning electron microscopy (SEM) image of the as-prepared MAIPb (methylammonium lead iodide perovskite).

In order to access the sorption behavior of these materials in aqueous phase, N_2 sorption can provide some useful information for the characterization and evaluation of the performance of the photocatalyst [48]. As indicated by the analysis in Figure S2, the sample showed type III according to IUPAC classification.

Figure S3 shows EDS spectra of the sample. The analysis confirms the presence of C, N, O, Pb, and I. The ratio C:N:I:Pb was found to be 4.06:0.58:42.33:49.62. Lower signals for carbon and nitrogen can be assigned to their lighter atomic weights. XRD pattern of the sample is presented in Figure 3A. Hexagonal crystal system was mainly detected with space groups *P3m1*. Dominant diffraction peaks at 2θ = 14.7, 30.2, 40, 46.26, and 53.01° were assigned respectively to the (002), (012), (104), (110), and (106) facets of the hexagonal crystalline structure. Peaks at 14.7 and 30.2° were also be indexed to (110) and (220) facets of the tetragonal structure of perovskite according to literature [49]. It should be noted the diffraction peaks of PbI_2, assigned at 2θ equals to 12.8°. The miller indexes (h, k, l) recorded suggested more than one preferred crystal orientation in our samples.

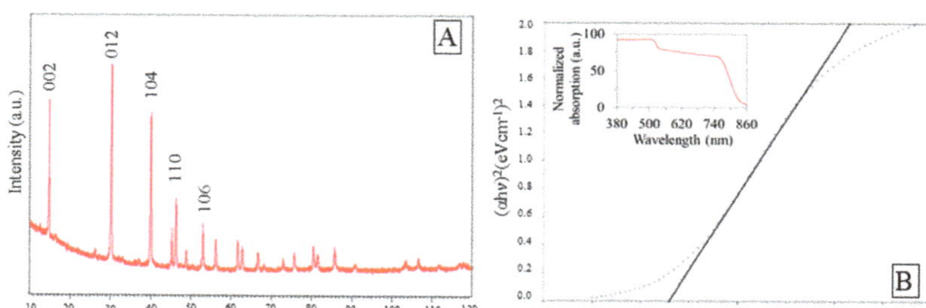

Figure 3. (**A**) X-ray diffraction (XRD) spectrum of the as-prepared MAIPb, (**B**) UV-vis spectrum and Tauc plot of the as prepared MAIPb.

Figure S4 displays FT-IR spectrum of the synthesized organo-halide perovskite. The sample showed broad vibrations N-H from 2800 to 3350 cm^{-1}, the characteristic features of hydrogen bonds overlapped the C-H vibrations signs. The peaks at 1450 cm^{-1} and around 650–750 cm^{-1} belong to the organic cation vibrations since the Pb-I and Pb-I-Pb appeared in very lower energy [50,51]. Peaks displayed at 1500 and 956 cm^{-1} an be respectively assigned to N-Pb-I stretching mode and Pb-I-NH bending. The wide bend around 3100 cm^{-1} was assigned to CH-NH stretching vibration [52].

The optical properties were further investigated in terms of light absorption capability because the absorption of light energy is one of the key of photocatalytic processes. Hybrid organic–inorganic perovskite achieves an optical absorbance across the entire visible spectrum as highlighted by Dualeh [53]. Carrier diffusion lengths was found to reach up to 100 nm for both electrons and holes in MAIPb via transient photo-luminescence measurements [54,55]. A nearly instantaneous charge generation and dissociations of balanced free charge carriers with high mobility has been observed, and the charges were proved to remain in that state for up to tens of microseconds [56]. From previous literature, it was found that the electronic levels for hybrid perovskites consist of an antibonding hybrid state between the Pb-s and I-p and a non-bonding hybrid state between the Pb-p and I-p orbitals corresponding to highest occupied and lowest unoccupied molecular orbitals, respectively [57]. The electronic properties were not influenced by organic fraction. In particular, Frost et al. showed that VB transition is primarily affected by the ionization potential of halogen ions contribution [58].

In Figure 3B the optical band gap of the perovskite was calculated. From extrapolation of the linear part of the Tauc plot (Kubelka–Munk theory), the optical gap was estimated to be 1.58 eV, which is in close agreement with previous reports [18,21,59–62].

XPS measurements were performed in order to investigate the chemical bonding states of the element in the envisaged catalyst Figure 4A. According to Navas et al. [18], peaks at 143 and 138.1 eV can be assigned to Pb 4f (Figure 5B); peaks around 412 eV, showed in Figure 4B, were assigned to Pb $4d_{5/2}$. The bigger peaks can be associated with the Pb component in the halide hybrid perovskite structure, while the smaller to metallic Pb probability decomposed from PbI during the synthesis [63]. Peak corresponding to 401 eV peak were assigned to N1s Figure 4D. In accordance with the studies performed by Chen et al. [64], N state may vary and the associated peaks can be found at different BE. Different peaks positions were found in a range of 396–404 eV in agreement with Nakamura and Mrowetz et al. [65]. Conforming to the study of Navas et al. [18], the peaks shown in Figure 4C belong to I $3d_{3/2}$ and I $3d_{5/2}$. It was further shown that the spectrum shows well separated spin–orbit components, separation of around 11.4–11.5 eV was recorded as typical evidence of the presence of I^{-0} [18]. Figure 4E shows peak belongs to C1s around 285 eV. Shen et al. [54], in their interesting research on hybrid organic-inorganic perovskite for solar cell application, assigned this peak to the methyl group. The conclusion obtained here agrees well with that reported by previous literature confirming the achievement of the synthesis processes [18,63–66].

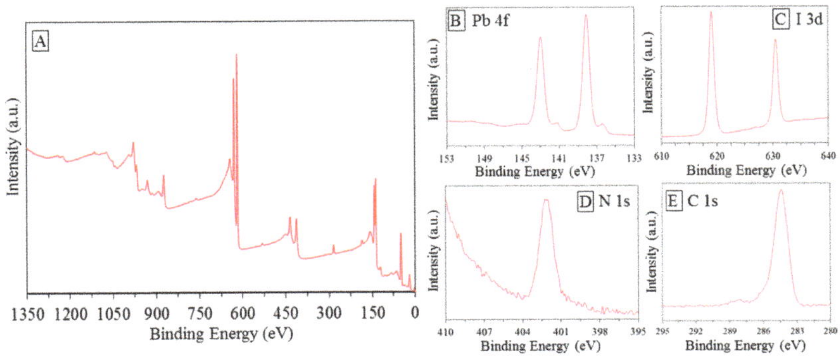

Figure 4. X-ray photoelectron spectroscopy (XPS) spectra of the as-prepared MAIPb: (**A**) general spectrum, (**B–E**) zooming on specific binding energy range.

3.2. Photocatalytic Activity

Among the persistent contaminants, organic dye molecules are toxic and their uncontrolled discharge from various industries into the water can have a huge impact on the environment [67]. In our study, the photocatalytic activity of the synthesized nano-catalyst was examined on RhB removal, which is considered as one of the most abundant dyes in the textile industries effluents and commonly chosen as model pollutant for photocatalytic treatment [68,69]. The photocatalytic performance of investigated material was evaluated as the decrease of the relative concentrations of RhB (C/C_0) plotted over time in different conditions. The removal efficiency achieved by photolysis was found to be negligible. This fact suggests that the chosen pollutant owns excellent photo-stability, as highlighted by Drexhage et al. and Beija et al. [70,71]. Control experiment in dark conditions was evaluated. The results showed moderate affinity between the halide perovskite and RhB molecules in terms of adsorption in darkness. The result are in accordance with the low surface area measured by BET analysis. As expected, significant improvement on RhB removal efficiency was observed during the photocatalytic experiments Figure 5A. After 3 h of irradiation, the concentration of RhB greatly decreased with respect to the initial concentration, proving the activity of the as-synthesized photocatalyst. The UV-vis spectra indicate that the main absorbance peak was reduced as a function of irradiation time and the dye molecules were decolorized. On the other hand, the peak position was found to be invariable and the diminishing intensity suggested that the fused aromatic ring structures and dye chromophores were destroyed (Figure S5). Kibombo et al. achieved similar results during their researches on optimization of photocatalysts for persistent organic pollutant remediation in wastewater management [72]. In their work it was deeply explained how the reactive oxygen species attack the auxochromic groups, induce N-de-ethylation of the alkyl amine group and how photogenerated holes can degrade both RhB suspended molecules and N-de-ethylated products. As depicted in Figure 5, the removal efficiency appeared at the very first interval (15 min), this is in accordance with the ROS generation that is higher at the earlier step of irradiation [73–75].

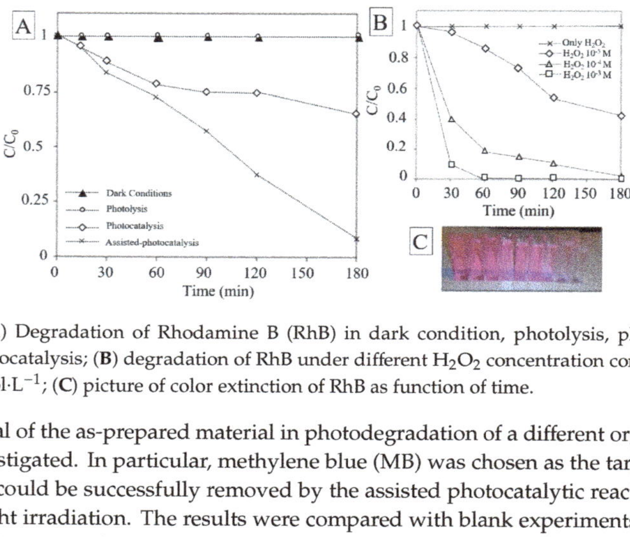

Figure 5. (**A**) Degradation of Rhodamine B (RhB) in dark condition, photolysis, photocatalysis, assisted-photocatalysis; (**B**) degradation of RhB under different H_2O_2 concentration conditions 10^{-4}, 10^{-3}, 10^{-5} mol·L^{-1}; (**C**) picture of color extinction of RhB as function of time.

The potential of the as-prepared material in photodegradation of a different organic compound was further investigated. In particular, methylene blue (MB) was chosen as the target contaminant. Methylene blue could be successfully removed by the assisted photocatalytic reaction after 60 min under visible-light irradiation. The results were compared with blank experiments to demonstrate the photocatalytic nature of the reaction. The results and the comparison are shown in Supporting Information (Figure S6).

The photocatalytic activity of the as-prepared nanoparticles showed higher photocatalytic efficiency for MB dye compared to RhB. The differences in the recorded efficiencies can be attributed to the chemical structures of the organic dyes and the nature of the functional groups present on their surfaces.

3.3. *Effect of H_2O_2 on the Photocatalysis Treatment*

Many techniques have been applied to reduce the effect of recombination of charges and to enhance the heterogenous photocatalysis performance. Among these techniques, the assistance of external electron acceptor such as hydrogen peroxide (H_2O_2) in the photocatalytic process has gained more and more attention. The effect of H_2O_2 on photocatalytic oxidation of RhB in aqueous suspensions of the as-synthesized material was investigated. Various concentrations of oxidant were used. Test without the presence of a photocatalyst was performed. In addition, the photocatalytic degradation of RhB was found to follow the pseudo first-order reaction model:

$$\ln(C/C_0) = -kt \tag{2}$$

The degradation rate constant k and the correlation coefficient of the curve R^2 were obtained using regression analysis. The value of R^2 were higher than 0.92, thus it was assumed that the regression line fits well with the data (Table 1).

Table 1. Degradation rate constant k and the correlation coefficient.

Experiment	Rate (s^{-1})	R^2
Assisted photocatalysis H_2O_2 10^{-5} mol·L^{-1}	0.0045	0.94
Assisted photocatalysis H_2O_2 10^{-3} mol·L^{-1}	0.0215	0.92
Assisted photocatalysis H_2O_2 10^{-3} mol·L^{-1}	0.1087	0.92

The reaction rate increased with H_2O_2 dosages. For the highest concentration of oxidant (10^{-3} mol·L^{-1}), the kinetic rate was found to be almost 25 times higher than the lowest concentration and 5 times higher than the average concentration. For practical application and considering the cost of hydrogen peroxide, 10^{-4} mol·L^{-1} was considered as the optimal value. The combination of halide perovskite and H_2O_2 under visible-light illumination was found to greatly enhance the degradation rates of RhB. When H_2O_2 concentration increases, more hydroxyl radicals are produced thus the oxidation rate increases. ROS were considered as dominant mechanism in the photocatalytic process. The first hypothesis is a direct photolysis of H_2O_2 by visible light that may generate free radicals at a wavelength of 405 nm [76]. A second minor mechanism proposed by Ollis et al. [77] and Ilisz et al. [78] suggested that H_2O_2 may partially contribute to the rate enhancement of photo-catalytic process behaving as an electron acceptor. According to these theories, H_2O_2 cannot only generate ·OH but also as electron acceptor, reduce the electrons-holes recombination increasing the photocatalytic efficiency. On the other hand Dionysiou et al. [79] in their studies on assisted-H_2O_2-photocatalysis showed that high concentrations of hydrogen peroxide may decrease the degradation rates because of the consumption of hydroxyl radicals.

3.4. Effect of Catalyst Loading

The effect of catalyst load on the ability to remove RhB in aqueous solution is shown in Figure 6A. The results suggest that the removal performance increased with the catalyst load up to 0.5 g·L^{-1} and decreased when the load is higher. This is in agreement with the case observed in heterogeneous photo-catalysis reaction. This behavior can be rationalized both in terms of availability of active sites on material surface and light penetration of photo-activating light into the system. The availability of active sites increased with catalyst loading, but on contrary the light penetration and, hence, the photo-activated volume of particles decreased [80]. Moreover, higher amount of catalyst may induce the deactivation of particles by collision with ground state molecules reducing the rate of reaction [81]. The trade-off of these effects was studied by considering also the organic contaminant concentration.

3.5. Effect of Initial Concentration of RhB

The effect of RhB initial concentration is an important parameter to consider [82]. Figure 6B depicts the effect of RhB initial concentration on its removal. The result reveals that the increase of the RhB concentration decreases the removal, corresponding to those from literature [1].

At higher RhB concentration, the generation of radicals on the surface of catalyst may be reduced by the competition of the active sites covered by RhB ions. Moreover, with the increase in the concentration the photons may be intercepted before they can reach the catalyst surface, decreasing the absorption of photons by the catalyst [83]. Higher concentration of RhB may also cause aggregation and even surface dimerization and have consequentially an effect on the degradation rates [42].

3.6. Effect of Initial pH

The pH of the dye solution was altered by adding incremental amounts of either dilute HCl or diluted NaOH in order to study the effect of pH on dye removal. Previously, it was confirmed that none of the salts used had any effect on the dye spectra in the absence of light. The solution was subjected to irradiation and change in absorbance value was noted.

The removal rate was found to increase in acidic media as shown in Figure 6C. The photolytic dye degradation appeared to be the best at pH 3 and it decreased when pH was increased. The results implied that in alkaline medium new oxidizing species, such as hydroperoxy anion can be formed. The new species can react with both the reactive oxygen species such as hydroxyl radicals as well as H_2O_2 molecules. This can consequently lower the dye contaminant removal rate. Future studies will be required to clarify the effect of pH on dye discoloration.

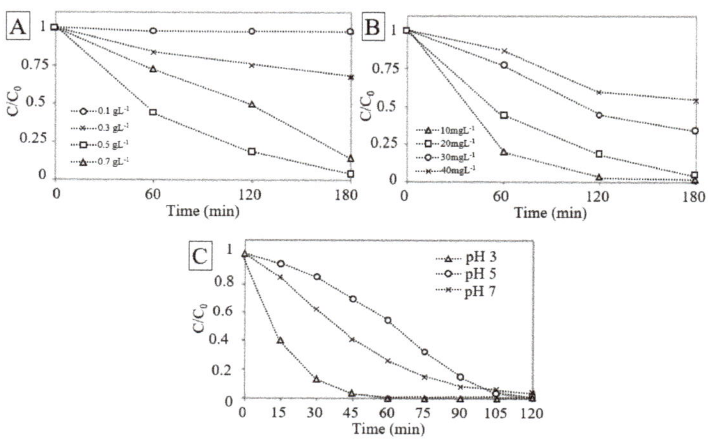

Figure 6. (**A**) Effect of catalyst load. (**B**) Effect of RhB initial concentration. (**C**) Effect of RhB initial pH value.

3.7. Effect of Temperature on H_2O_2-Assisted Photo-Catalysis

According to Wang et al. [84], temperature is another parameter that affects the heterogenous photo-catalysis. Therefore, in this study, 25 °C, 35 °C, 45 °C were selected to examine the effect of temperature on photo-catalysis under visible light irradiation. As the temperature increased from 25 to 45 °C, the first-order rate constant k1 increased almost 40% (Table 2). This behavior was associated to a decrease in the viscosity and to an enhanced diffusion of the sorbate molecule [85].

Table 2. Impact of temperature on the RhB removal kinetic rate under the $CH_3NH_3PbI_3$/visible irradiation system, experimental conditions RhB: (20 mg·L^{-1}), H_2O_2 (10^{-4} M), photo-catalyst (0.5 g·L^{-1}), pH 5.

Temperature (°C)	Kinetic rate (min^{-1})	R^2
25	0.0328	0.9632
35	0.0499	0.9143
45	0.0840	0.8526

The Arrhenius equation was used to determine the activation energy as follows:

$$K = A^{*}\exp(-Ea/RT) \tag{3}$$

where K is the constant rate that controls the entire process, A is the Arrhenius constant, T the solution temperature in K, Ea apparent activation energy ($kJ \cdot mol^{-1}$), and R the ideal gas constant 0.0083 kJ $mol^{-1} \cdot K^{-1}$. The data are fitted using a linear regression (R^2 0.9935). From the Arrhenius-type plot (Figure 7). Ea value was calculated as 36.96 $kJ \cdot mol^{-1}$. Mcheik and El Jamal found similar result in their study on removal of RhB with persulfate and iron activation [86]. The reaction appeared to be activated also at room temperature and proceeded with relatively low energy barrier.

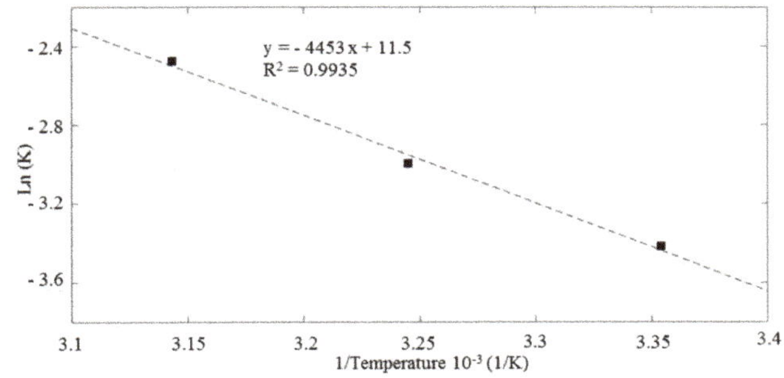

Figure 7. Arrhenius-type plot for the evaluation of the activation energy of the reaction.

3.8. Recyclability of the H_2O_2-Assisted Photo-Catalysis System

Recyclability of the photocatalyst represents one of the most important advantages of a heterogeneous application. Thus, the recyclability of the synthesized material was evaluated by using H_2O_2 to activate the process for multiple cycles. Figure 8 shows three cycles of the RhB removal using the H_2O_2-hybrid organic-inorganic perovskite system irradiated under visible light. It can be seen that after 3 cycles, the system showed a stable and effective catalytic activity on the removal of the selected dye, and the activity loss was negligible. RhB degradation efficiency showed slight decrease from 93% to 80% after 120 min of the third treatment. The results obtained may be caused by active sites saturation. Moreover, the recycle was performed in series, thus a slight decrease in photocatalyst content should be considered. It must be mentioned that the main aim of the former study is to investigate the potential of HOIPs in photocatalytic processes. Further development on material and process technology should be applied.

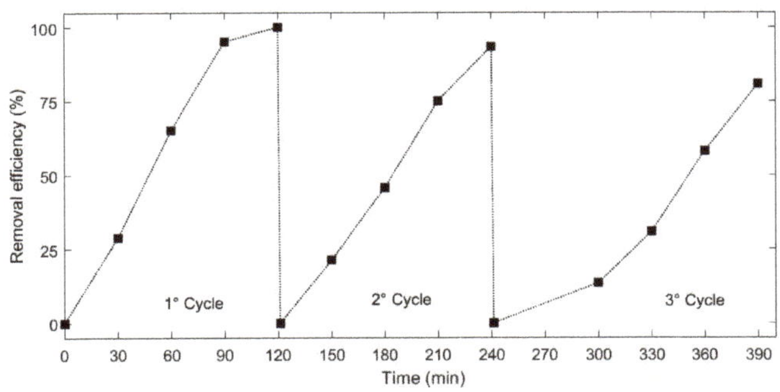

Figure 8. Degradation profile of RhB under assisted visible light photo-catalysis for three cycles.

3.9. Active Species and Possible Mechanism

In a typical photocatalytic application, when a semiconductor is irradiated with equivalent or greater light-energy, the electrons (e^-) in the valence band (VB) are excited into the conduction band (CB) leaving holes (h^+) in the VB. The photo-generated electrons and holes trigger the redox reaction.

When the bottoms of the CB is below the reduction potential of H^+ to H_2 (0 V vs. NHE), and the tops of the VB must be located more positively than the oxidation potential of H_2O to O_2 (1.23 V vs. NHE) both oxidation and reduction sites are created [87]. The electron/hole pairs and reactive oxygen species (ROS), including $O_2{\cdot}^-$, and ·OH, are widely considered the main active species responsible for photocatalytic degradation of organic contaminants [88,89].

As deeply studied by Han et al. [90], the electron spin resonance (ESR) spin-trap technique confirms the presence of free radicals. DMPO and TEMP were used as spin trap for superoxide or hydroxide radicals anions ($O_2{\cdot}^-$, ·OH) and singlet oxygen species ($^1O_2$), respectively.

In detail, upon visible light photo-excitation of the mixture of the organo-halide perovskite and diamagnetic 2,2,6,6-tetramethylpiperidine (TEMP), three lines with equal intensities were observed in the recorded spectrum in Figure 9. This indicates the capture of singlet oxygen ($^1O_2$) generated by TEMP, leading to the formation of the TEMPO radical. The irradiation period was set at 5 min, a signal of g = 2.0001 appeared confirming photo-generation of radicals. The time of irradiation increased and the intensity of peaks decreased, after half-hour of irradiation the resulting spectrum is shown in Figure 9A. The decrease in spectrum intensity of peaks suggests that $^1O_2$ radical generation occurred in the very first intervals of the photo-catalytic process that is mainly due to their nano-second lifetime [46,91,92]. 5,5-dimethylpyrroline N-oxide (DMPO) was utilized as superoxide and hydroxide radical anions ($O_2{\cdot}^-$, ·OH) spin trap. Four typical peaks appeared in the ESR spectrum revealing the presence of the radicals, g factor was found equal to 1.9985. Later, the sampling period was increased, and the lower peaks were recorded, indicating that radical generation belongs to the initial period of irradiation. Figure 9B shows the radical peaks after 5 min of irradiation. The signal recorded after 30 min shows a decrease in the intensity of peaks implying that no more radicals are present in the solution.

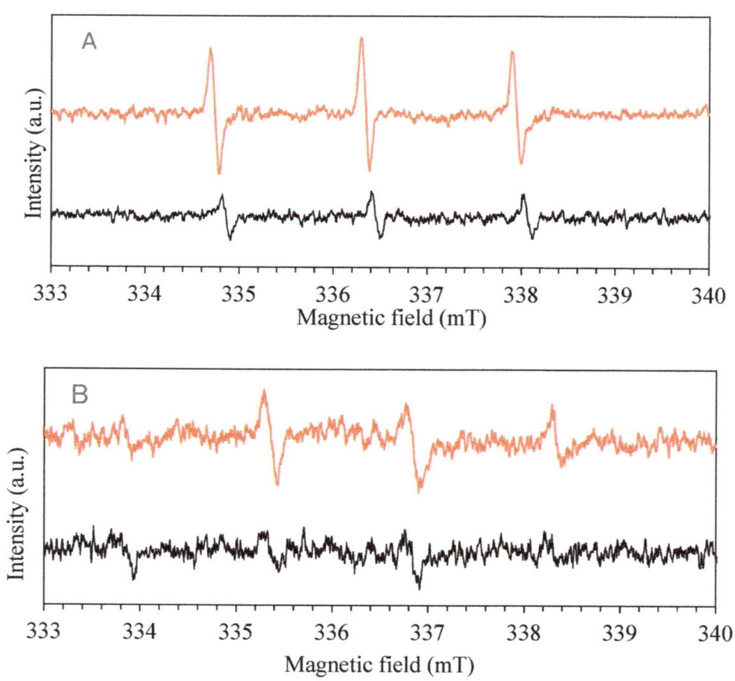

Figure 9. Electron paramagnetic resonance (EPR) spectra using as spin-trap: (**A**) TEMP, red for 5 min, black for 30 min; (**B**) DMPO, red for 5 min, black for 30 min.

Finally, to evaluate also the effect of RhB in the production of radicals, a solution of equal content (100 μL) of RhB (20 mg·L^{-1}) and DMPO (100 mM) was prepared and irradiated in the presence of photo-catalyst. After an irradiation time of 5 min, the ESR spectrum was recorded revealing the presence of ·OH radicals (Figure 10). Four typical peaks were recorded also in presence of RhB, revealing a potential synergetic effect between photo-catalyst and organic dye in the production of hydroxyl and superoxide species. A mixed solution of RhB and DMPO was also prepared in the absence of photo-catalyst to confirm the absence of the radicals.

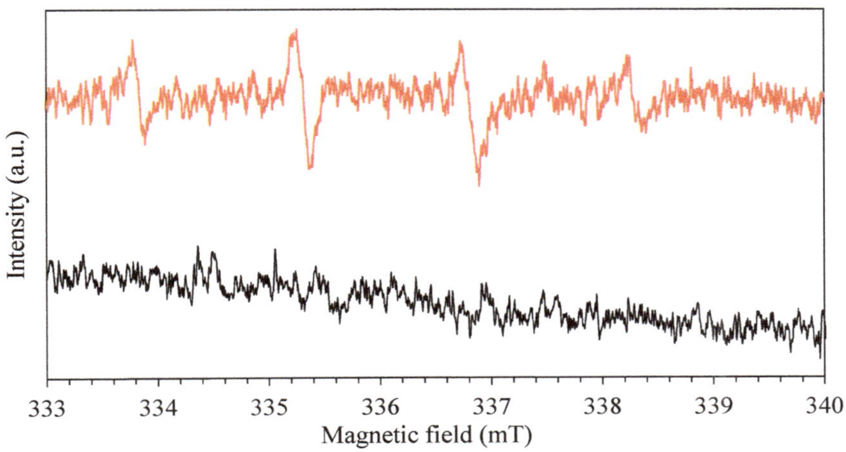

Figure 10. EPR spectra using DMPO as spin-trap in presence of RhB, red with catalyst and black without catalyst.

From the results described above, it may be concluded that both $^1O_2$, $O_2{\cdot}^-$ and ·OH radicals were produced during the visible-light photo-catalyst treatment of RhB [93,94].

The photocatalytic degradation process proceeds through excitation, transportation, and degradation pathways. As highlighted by Yin et al. [19], during the investigation mechanism of photocatalytic degradation of RhB by TiO_2/Eosin-Y system under visible light, dye molecules transfer electrons onto conduction bands (CB) of catalyst leading to the formation of dye cationic radicals. Then the involved electrons generate a series of active oxygen species such as $O_2{\cdot}^-$, ·OH, and $^1O_2$ which are considered to be involved in the organic contaminant degradation. In a similar study performed by Dutta et al. [95], two main mechanisms were proposed to promote dye degradation, one governed by dye sensitization and the other by the photo-catalyst excitation. In the self-sensitized dye degradation, the photo-induced electrons flow from the dyes to photo-catalyst surface as suggested by their potential energy values. In particular, Lv et al. [96], with their respective co-authors, deeply described the direction of the charge flow; the difference in the potential energy between the CBs induces the electrons to transfer from higher energy level of the photo-excited dye to the lower ones of catalyst. On the other hand, visible light excitation of MAIPb structures could also generate holes in the valance band (VB) and electrons in the CB. Egger et al. studied the tunability of VB (ionization potential) and CB (electron affinity) energies by the atomic orbitals of the anions and cations in different organohalide perovskite [97]. Band energy and band gap engineering of these organic–inorganic solids are indeed possible to be controlled by the chemical composition, and iodine presence was found to upshift the VB and generally narrowing the band gap, favorable condition for bleaching organic compound in aqueous solutions.

The CB transported electrons in both the materials may react with the dissolved oxygen in the water to produce a reactive oxygen species, main responsible for the oxidative dye degradation under visible light irradiation. As confirmed by an interesting study on nanosized Bi_2WO_6 performed under visible light by Fu et al. [98], the presence of oxygen is responsible for the activation of photo-catalysis process. In their experiments, they confirmed the importance of the presence of dissolved oxygen in the treated solution, since its effect is primarily to act as an efficient e^- trap, leading to the generation of reactive oxygen species and preventing the recombination of charges. Furthermore, Dutta et al. highlighted a similar conclusion in their study on ternary nano-composite based on cadium sulphide (CdS), TiO_2, and graphene oxide. Herein, they proved how generated electrons react with the dissolved oxygen in water to produce a reactive oxidizing agent initially in the specific form of oxygen radical anion $O_2\cdot^-$, responsible for the oxidative dye degradation under visible light irradiation [95].

Based on the previous discussion, a possible mechanism of RhB is depicted in Figure 11. After self-sensitization of RhB and the excitation of organohalide perovskite, separation of charges occurs, and transport of electron is promoted. On the other hand, dissolved oxygen can act as an electron acceptor, and can be reduced by the promoted electron in the conduction band to form a superoxide specie $O_2\cdot^-$ (3). The $O_2\cdot^-$ can subsequently re-oxidize to $^1O_2$ or, in the presence of water and H_2O_2, it can form ·OH. The strong oxidation power of the hole enables a one-electron oxidation step with water to produce a hydroxyl radical ·OH. These radicals are highly ROS, able to oxidize directly organic contaminant. In our study, the generation of $O_2\cdot^-$ and ·OH was confirmed by the ESR spectra by using DMPO as the spin trap reagent [14], instead, TEMP was used to detect singlet oxygen and it proved electrons and holes generation during visible light irradiation [46].

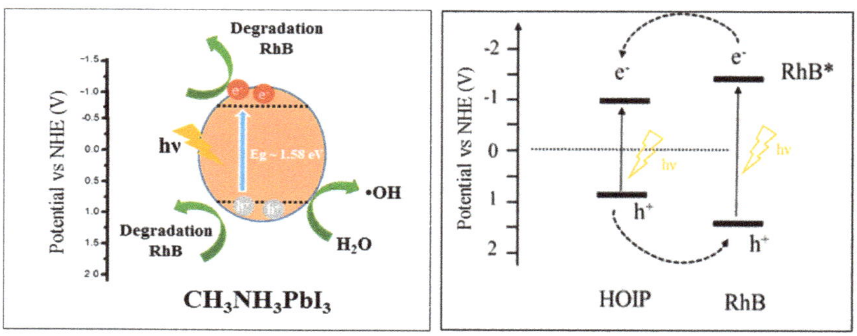

Figure 11. Proposed band gap energy diagram and charge transfer between RhB and photo-catalyst.

4. Conclusions

In conclusion, bare MAIPb were easily synthesized for the photocatalytic degradation of organic dye pollutants. The degradation performance study suggested that RHB was completely degraded after 180 min of treatment assisted by H_2O_2-MAIPb system under visible light irradiation. In this work, we have shown that the outstanding optoelectronic properties of MAIPb can be addressed for photocatalytic degradation of organic compounds. The results constitute a significant step forward in the application of hybrid halide perovskite for solar-driven catalytic processes. It is important to mention that the systematic evaluation of the environmental conditions must be deeply studied.

Supplementary Materials: The supplementary materials are available online at http://www.mdpi.com/2079-4991/10/1/115/s1.

Author Contributions: B.-M.B. as author of research and editor of manuscript, S.B.H. as co-supervisor, M.S. as supervisor and funding provider. All authors have read and agreed to the published version of the manuscript.

Funding: This research was funded by Maa-ja vesitekniikan tuki Foundation.

Conflicts of Interest: The authors declare no conflict of interest.

References

1. Chakrabarti, S.; Dutta, B.K. Photocatalytic degradation of model textile dyes in wastewater using ZnO as semiconductor catalyst. *J. Hazard. Mater.* **2004**, *112*, 269–278. [CrossRef] [PubMed]
2. Crake, A.; Christoforidis, K.C.; Godin, R.; Moss, B.; Kafizas, A.; Zafeiratos, S.; Durrant, J.R.; Petit, C. Titanium dioxide/carbon nitride nanosheet nanocomposites for gas phase CO_2 photoreduction under UV-visible irradiation. *Appl. Catal. B Environ.* **2019**, *242*, 369–378. [CrossRef]
3. Fujishima, A.; Zhang, X.; Tryk, D.A. Heterogeneous photocatalysis: From water photolysis to applications in environmental cleanup. *Int. J. Hydrog. Energy* **2007**, *32*, 2664–2672. [CrossRef]
4. Hoffmann, M.R.; Martin, S.T.; Choi, W.; Bahnemann, D.W. Environmental Applications of Semiconductor Photocatalysis. *Chem. Rev.* **1995**, *95*, 69. [CrossRef]
5. Huang, J.H.; Huang, K.L.; Liu, S.Q.; Wang, A.T.; Yan, C. Adsorption of Rhodamine B and methyl orange on a hypercrosslinked polymeric adsorbent in aqueous solution. *Colloids Surfaces A Physicochem. Eng. Asp.* **2008**, *330*, 55–61. [CrossRef]
6. Lefebvre, O.; Moletta, R. Treatment of organic pollution in industrial saline wastewater: A literature review. *Water Res.* **2006**, *40*, 3671–3682. [CrossRef]
7. Teng, F.; Liu, Z.; Zhang, A.; Li, M. Photocatalytic Performances of Ag_3PO_4Polypods for Degradation of Dye Pollutant under Natural Indoor Weak Light Irradiation. *Environ. Sci. Technol.* **2015**, *49*, 9489–9494. [CrossRef]
8. Spasiano, D.; Marotta, R.; Malato, S.; Fernandez-Ibañez, P.; di Somma, I. Solar photocatalysis: Materials, reactors, some commercial, and pre-industrialized applications. A comprehensive approach. *Appl. Catal. B Environ.* **2015**, *170–171*, 90–123. [CrossRef]
9. Patrocinio, A.O.T.; Schneider, J.; França, M.D.; Santos, L.M.; Caixeta, B.P.; Machado, A.E.H.; Bahnemann, D.W. Charge carrier dynamics and photocatalytic behavior of TiO_2 nanopowders submitted to hydrothermal or conventional heat treatment. *RSC Adv.* **2015**, *5*, 70536–70545. [CrossRef]
10. Natarajan, T.S.; Bajaj, H.C.; Tayade, R.J. Enhanced direct sunlight photocatalytic oxidation of methanol using nanocrystalline TiO_2 calcined at different temperature. *J. Nanopart. Res.* **2014**, *16*, 2713. [CrossRef]
11. Kou, J.; Lu, C.; Wang, J.; Chen, Y.; Xu, Z.; Varma, R.S. Selectivity Enhancement in Heterogeneous Photocatalytic Transformations. *Chem. Rev.* **2017**, *117*, 1445–1514. [CrossRef] [PubMed]
12. Zhang, X.; Wang, Y.; Hou, F.; Li, H.; Yang, Y.; Zhang, X.; Yang, Y.; Wang, Y. Effects of Ag loading on structural and photocatalytic properties of flower-like ZnO microspheres. *Appl. Surf. Sci.* **2017**, *391*, 476–483. [CrossRef]
13. Zhang, X.; Yang, Y.; Huang, W.; Yang, Y.; Wang, Y.; He, C.; Liu, N.; Wu, M.; Tang, L. g-C 3 N 4 /UiO-66 nanohybrids with enhanced photocatalytic activities for the oxidation of dye under visible light irradiation. *Mater. Res. Bull.* **2018**, *99*, 349–358. [CrossRef]
14. Geng, X.; Li, W.; Xiao, F.; Wang, D.; Yang, L. Effect of in situ Fe(II)/Fe(III)-doping on the visible light-Fenton-like catalytic activity of Bi/BiOBr hierarchical microspheres. *Catal. Sci. Technol.* **2017**, *7*, 658–667. [CrossRef]
15. Takanabe, K. Photocatalytic Water Splitting: Quantitative Approaches toward Photocatalyst by Design. *ACS Catal.* **2017**, *7*, 8006–8022. [CrossRef]
16. Kojima, A.; Teshima, K.; Shirai, Y.; Miyasaka, T. Organometal Halide Perovskites as Visible-Light Sensitizers for Photovoltaic Cells. *J. Am. Chem. Soc.* **2009**, *131*, 6050–6051. [CrossRef]
17. Tang, H.; He, S.; Peng, C. A Short Progress Report on High-Efficiency Perovskite Solar Cells. *Nanoscale Res. Lett.* **2017**, *12*, 410. [CrossRef]
18. Navas, J.; Sánchez-Coronilla, A.; Gallardo, J.J.; Hernández, N.C.; Piñero, J.C.; Alcántara, R.; Fernández-Lorenzo, C.; de los Santos, D.M.; Aguilar, T.; Martín-Calleja, J. New insights into organic–inorganic hybrid perovskite $CH_3NH_3PbI_3$ nanoparticles. An experimental and theoretical study of doping in Pb^{2+} sites with Sn^{2+}, Sr^{2+}, Cd^{2+} and Ca^{2+}. *Nanoscale* **2015**, *7*, 6216–6229. [CrossRef]
19. Yin, W.J.; Shi, T.; Yan, Y. Unique properties of halide perovskites as possible origins of the superior solar cell performance. *Adv. Mater.* **2014**, *26*, 4653–4658. [CrossRef]
20. Hantusch, M.; Bessergenev, V.; Mateus, M.C.; Knupfer, M.; Burkel, E. Electronic properties of photocatalytic improved Degussa P25 titanium dioxide powder. *Catal. Today* **2018**, *307*, 111–118. [CrossRef]
21. Kim, H.-S.; Lee, C.-R.; Im, J.-H.; Lee, K.-B.; Moehl, T.; Marchioro, A.; Moon, S.-J.; Humphry-Baker, R.; Yum, J.-H.; Moser, J.E.; et al. Lead Iodide Perovskite Sensitized All-Solid-State Submicron Thin Film Mesoscopic Solar Cell with Efficiency Exceeding 9%. *Sci. Rep.* **2012**, *2*, 591. [CrossRef] [PubMed]
22. Mauro, J.C. Topological constraint theory of glass. *Am. Ceram. Soc. Bull.* **2011**, *90*, 31–37. [CrossRef]

23. Stranks, S.D.; Stranks, S.D.; Eperon, G.E.; Grancini, G.; Menelaou, C.; Alcocer, M.J.P.; Leijtens, T.; Herz, L.M.; Petrozza, A.; Snaith, H.J. Electron-Hole Diffusion Lengths Exceeding. *Science* **2014**, *342*, 341–344. [CrossRef] [PubMed]
24. Chen, Q.; de Marco, N.; Yang, Y.; Song, T.B.; Chen, C.C.; Zhao, H.; Hong, Z.; Zhou, H.; Yang, Y. Under the spotlight: The organic-inorganic hybrid halide perovskite for optoelectronic applications. *Nano Today* **2015**, *10*, 355–396. [CrossRef]
25. Xing, G.; Mathews, N.; Sun, S.; Lim, S.S.; Lam, Y.M.; Gratzel, M.; Mhaisalkar, S.; Sum, T.C. Long-Range Balanced Electron- and Hole-Transport Lengths in Organic-Inorganic $CH_3NH_3PbI_3$. *Science* **2013**, *342*, 344–347. [CrossRef] [PubMed]
26. Singh, T.; Kulkarni, A.; Ikegami, M.; Miyasaka, T. Effect of Electron Transporting Layer on Bismuth-Based Lead-Free Perovskite (CH3NH3)3 Bi_2I_9 for Photovoltaic Applications. *ACS Appl. Mater. Interfaces* **2016**, *8*, 14542–14547. [CrossRef]
27. Butler, K.T.; Frost, J.M.; Walsh, A. Band alignment of the hybrid halide perovskites $CH_3NH_3PbCl_3$, $CH_3NH_3PbBr_3$ and $CH_3NH_3PbI_3$. *Mater. Horiz.* **2015**, *2*, 228–231. [CrossRef]
28. Jeon, N.J.; Noh, J.H.; Yang, W.S.; Kim, Y.C.; Ryu, S.; Seo, J.; Seok, S.I. Compositional engineering of perovskite materials for high-performance solar cells. *Nature* **2015**, *517*, 476–480. [CrossRef]
29. Zhou, H.; Shen, Q.; Li, G.; Luo, S.; Song, T.; Duan, H.-S.; Hong, Z.; You, J.; Liu, Y.; Yang, Y. Interface engineering of highly efficient perovskite solar cells. *Science* **2014**, *345*, 542–546. [CrossRef]
30. Liu, M.; Johnston, M.B.; Snaith, H.J. Efficient planar heterojunction perovskite solar cells by vapour deposition. *Nature* **2013**, *501*, 395–398. [CrossRef]
31. Burschka, J.; Pellet, N.; Moon, S.-J.; Humphry-Baker, R.; Gao, P.; Nazeeruddin, M.K.; Grätzel, M. Sequential deposition as a route to high-performance perovskite-sensitized solar cells. *Nature* **2013**, *499*, 316–319. [CrossRef] [PubMed]
32. Niu, G.; Guo, X.; Wang, L. Review of recent progress in chemical stability of perovskite solar cells. *J. Mater. Chem. A* **2015**, *3*, 8970–8980. [CrossRef]
33. Acik, M.; Darling, S.B. Graphene in perovskite solar cells: Device design, characterization and implementation. *J. Mater. Chem. A* **2016**, *4*, 6185–6235. [CrossRef]
34. Gupta, A.K.; Pal, A.; Sahoo, C. Photocatalytic degradation of a mixture of Crystal Violet (Basic Violet 3) and Methyl Red dye in aqueous suspensions using Ag^+ doped TiO_2, Dye. *Pigment* **2006**, *69*, 224–232. [CrossRef]
35. Girija, K.; Thirumalairajan, S.; Mastelaro, V.R.; Mangalaraj, D. Photocatalytic degradation of organic pollutants by shape selective synthesis of β-Ga_2O_3 microspheres constituted by nanospheres for environmental remediation. *J. Mater. Chem. A* **2015**, *3*, 2617–2627. [CrossRef]
36. McKinnon, N.K.; Reeves, D.C.; Akabas, M.H. 5-HT3 receptor ion size selectivity is a property of the transmembrane channel, not the cytoplasmic vestibule portals. *J. Gen. Physiol.* **2011**, *138*, 453–466. [CrossRef]
37. Im, J.-H.; Chung, J.; Kim, S.-J.; Park, N.-G. Synthesis, structure, and photovoltaic property of a nanocrystalline 2H perovskite-type novel sensitizer $(CH_3CH_2NH_3)PbI_3$. *Nanoscale Res. Lett.* **2012**, *7*, 353. [CrossRef]
38. Umari, P.; Mosconi, E.; de Angelis, F. Relativistic GW calculations on $CH_3NH_3PbI_3$ and $CH_3NH_3SnI_3$ Perovskites for Solar Cell Applications. *Sci. Rep.* **2015**, *4*, 4467. [CrossRef]
39. Green, M.A.; Ho-Baillie, A.; Snaith, H.J. The emergence of perovskite solar cells. *Nat. Photonics* **2014**, *8*, 506–514. [CrossRef]
40. Bernal, C.; Yang, K. First-principles hybrid functional study of the organic-inorganic perovskites $CH_3NH_3SnBr_3$ and $CH_3NH_3SnI_3$. *J. Phys. Chem. C* **2014**, *118*, 24383–24388. [CrossRef]
41. Noel, N.K.; Stranks, S.D.; Abate, A.; Wehrenfennig, C.; Guarnera, S.; Haghighirad, A.-A.; Sadhanala, A.; Eperon, G.E.; Pathak, S.K.; Johnston, M.B.; et al. Lead-free organic–inorganic tin halide perovskites for photovoltaic applications. *Energy Environ. Sci.* **2014**, *7*, 3061–3068. [CrossRef]
42. Rochkind, M.; Pasternak, S.; Paz, Y. Using dyes for evaluating photocatalytic properties: A critical review. *Molecules* **2015**, *20*, 88–110. [CrossRef] [PubMed]
43. Liu, N.; Huang, W.; Tang, M.; Yin, C.; Gao, B.; Li, Z.; Tang, L.; Lei, J.; Cui, L.; Zhang, X. In-situ fabrication of needle-shaped MIL-53(Fe) with 1T-MoS_2 and study on its enhanced photocatalytic mechanism of ibuprofen. *Chem. Eng. J.* **2019**, *359*, 254–264. [CrossRef]
44. Liu, N.; Huang, W.; Zhang, X.; Tang, L.; Wang, L.; Wang, Y.; Wu, M. Ultrathin graphene oxide encapsulated in uniform MIL-88A(Fe) for enhanced visible light-driven photodegradation of RhB. *Appl. Catal. B Environ.* **2018**, *221*, 119–128. [CrossRef]

45. Dvoranová, D.; Barbieriková, Z.; Brezová, V. Radical intermediates in photoinduced reactions on TiO_2 (An EPR spin trapping study). *Molecules* **2014**, *19*, 17279–17304. [CrossRef]
46. Li, Z.; Han, Y.; Gao, Z.; Wang, F. Supramolecular Engineering of Discrete Pt(II)···Pt(II) Interactions for Visible-Light Photocatalysis. *ACS Catal.* **2017**, *7*, 4676–4681. [CrossRef]
47. Li, D.; Wang, G.; Cheng, H.-C.; Chen, C.-Y.; Wu, H.; Liu, Y.; Huang, Y.; Duan, X. Size-dependent phase transition in methylammonium lead iodide perovskite microplate crystals. *Nat. Commun.* **2016**, *7*. [CrossRef]
48. Zhang, X.; Yang, Y.; Lv, X.; Wang, Y.; Liu, N.; Chen, D.; Cui, L. Adsorption/desorption kinetics and breakthrough of gaseous toluene for modified microporous-mesoporous UiO-66 metal organic framework. *J. Hazard. Mater.* **2019**, *366*, 140–150. [CrossRef]
49. Grabowska, E. Selected perovskite oxides: Characterization, preparation and photocatalytic properties-A review. *Appl. Catal. B Environ.* **2016**, *186*, 97–126. [CrossRef]
50. Pérez-Osorio, M.A.; Milot, R.L.; Filip, M.R.; Patel, J.B.; Herz, L.M.; Johnston, M.B.; Giustino, F. Vibrational Properties of the Organic-Inorganic Halide Perovskite $CH_3NH_3PbI_3$ from Theory and Experiment: Factor Group Analysis, First-Principles Calculations, and Low-Temperature Infrared Spectra. *J. Phys. Chem. C* **2015**, *119*, 25703–25718. [CrossRef]
51. Que, C.J.; Mo, C.J.; Li, Z.Q.; Zhang, G.L.; Zhu, Q.Y.; Dai, J. Perovskite-Like Organic-Inorganic Hybrid Lead Iodide with a Large Organic Cation Incorporated within the Layers. *Inorg. Chem.* **2017**, *56*, 2467–2472. [CrossRef] [PubMed]
52. Kumar, V.B.; Gouda, L.; Porat, Z.; Gedanken, A. Sonochemical synthesis of $CH_3NH_3PbI_3$perovskite ultrafine nanocrystal sensitizers for solar energy applications. *Ultrason. Sonochem.* **2016**, *32*, 54–59. [CrossRef] [PubMed]
53. Dualeh, A.; Moehl, T.; Tétreault, N.; Teuscher, J.; Gao, P.; Nazeeruddin, M.K.; Grätzel, M. Impedance spectroscopic analysis of lead iodide perovskite-sensitized solid-state solar cells. *ACS Nano* **2014**, *8*, 362–373. [CrossRef] [PubMed]
54. Shen, P.-S.; Chiang, Y.-H.; Li, M.-H.; Guo, T.-F.; Chen, P. Research Update: Hybrid organic-inorganic perovskite (HOIP) thin films and solar cells by vapor phase reaction. *APL Mater.* **2016**, *4*, 91509. [CrossRef]
55. Liu, J.; Liu, Y.; Liu, N.; Han, Y.; Zhang, X.; Huang, H.; Lifshitz, Y.; Lee, S.; Zhong, J.; Kang, Z. Metal-free efficient photocatalyst for stable visible water splitting via a two-electron pathway. *Science* **2015**, *6709*, 1–6.
56. Ponseca, C.S.; Savenije, T.J.; Abdellah, M.; Zheng, K.; Yartsev, A.; Pascher, T.; Harlang, T.; Chabera, P.; Pullerits, T.; Stepanov, A.; et al. Organometal halide perovskite solar cell materials rationalized: Ultrafast charge generation, high and microsecond-long balanced mobilities, and slow recombination. *J. Am. Chem. Soc.* **2014**, *136*, 5189–5192. [CrossRef]
57. Brivio, F.; Walker, A.B.; Walsh, A. Structural and electronic properties of hybrid perovskites for high-efficiency thin-film photovoltaics from first-principles. *APL Mater.* **2013**, *1*, 9–14. [CrossRef]
58. Frost, J.M.; Butler, K.T.; Brivio, F.; Hendon, C.H.; van Schilfgaarde, M.; Walsh, A. Atomistic origins of high-performance in hybrid halide perovskite solar cells. *Nano Lett.* **2014**, *14*, 2584–2590. [CrossRef]
59. Chatterjee, S.; Pal, A.J. Introducing Cu_2O Thin Films as a Hole-Transport Layer in Efficient Planar Perovskite Solar Cell Structures. *J. Phys. Chem. C* **2016**, *120*, 1428–1437. [CrossRef]
60. Zhao, Y.; Zhu, K. Charge Transport and Recombination in Perovskite $(CH_3NH_3)PbI_3$ Sensitized TiO_2 Solar Cells. *J. Phys. Chem. Lett.* **2013**, 2880–2884. [CrossRef]
61. Noh, J.H.; Im, S.H.; Heo, J.H.; Mandal, T.N.; Seok, S.I. Chemical Management for Colorful, Efficient, and Stable Inorganic—Organic Hybrid Nanostructured Solar Cells. *Nano Lett.* **2013**, *13*, 1764–1769. [CrossRef] [PubMed]
62. Zhao, Y.; Zhu, K. Optical bleaching of perovskite $(CH_3NH_3)PbI_3$ through room-temperature phase transformation induced by ammonia. *Chem. Commun.* **2014**, *50*, 1605–1607. [CrossRef] [PubMed]
63. Xie, H.; Liu, X.; Lyu, L.; Niu, D.; Wang, Q.; Huang, J.; Gao, Y. Effects of Precursor Ratios and Annealing on Electronic Structure and Surface Composition of $CH_3NH_3PbI_3$ Perovskite Films. *J. Phys. Chem. C* **2015**. [CrossRef]
64. Chen, J.; He, Z.; Li, G.; An, T.; Shi, H.; Li, Y. Visible-light-enhanced photothermocatalytic activity of ABO_3-type perovskites for the decontamination of gaseous styrene. *Appl. Catal. B Environ.* **2017**, *209*, 146–154. [CrossRef]
65. Mrowetz, M.; Balcerski, W.; Colussi, A.J.; Hoffmann, M.R. Oxidative power of nitrogen-doped TiO_2 photocatalysts under visible illumination. *J. Phys. Chem. B* **2004**, *108*, 17269–17273. [CrossRef]

66. Nakamura, R.; Tanaka, T.; Nakato, Y. Mechanism for visible light responses in anodic photocurrents at N-doped TiO_2 film electrodes. *J. Phys. Chem. B* **2004**, *108*, 10617–10620. [CrossRef]
67. Dutta, S.; Sarkar, S.; Ray, C.; Pal, T. Benzoin derived reduced graphene oxide (rGO) and its nanocomposite: Application in dye removal and peroxidase-like activity. *RSC Adv.* **2013**, *3*, 21475. [CrossRef]
68. Sinha, A.K.; Pradhan, M.; Sarkar, S.; Pal, T. Large-scale solid-state synthesis of Sn-SnO_2 nanoparticles from layered SnO by sunlight: A material for dye degradation in water by photocatalytic reaction. *Environ. Sci. Technol.* **2013**, *47*, 2339–2345. [CrossRef]
69. Ray, C.; Dutta, S.; Sarkar, S.; Sahoo, R.; Roy, A.; Pal, T. A facile synthesis of 1D nano structured selenium and Au decorated nano selenium: Catalysts for the clock reaction. *RSC Adv.* **2013**, *3*, 24313. [CrossRef]
70. Drexhage, K.H.H. Fluorescence efficiency of laser dyes. *J. Res. Natl. Bur. Stand. Sect. A Phys. Chem.* **1976**, *80*, 421. [CrossRef]
71. Beija, M.; Afonso, C.A.M.; Martinho, J.M.G. Synthesis and applications of Rhodamine derivatives as fluorescent probes. *Chem. Soc. Rev.* **2009**, *38*, 2410. [CrossRef] [PubMed]
72. Kibombo, H.S.; Rasalingam, S.; Koodali, R.T. Facile template free method for textural property modulation that enhances adsorption and photocatalytic activity of aperiodic titania supported silica materials. *Appl. Catal. B Environ.* **2013**, *142*, 119–128. [CrossRef]
73. Chen, H.; Xu, X. Ruddlesden-Popper compounds in the double-perovskite family $Sr_2FeTaO_6(SrO)_n$ (n = 0, 1 and 2) and their photocatalytic properties. *Appl. Catal. B Environ.* **2017**, *206*, 35–43. [CrossRef]
74. Xin, B.; Jing, L.; Ren, Z.; Wang, B.; Fu, H. Effects of simultaneously doped and deposited Ag on the photocatalytic activity and surface states of TiO_2. *J. Phys. Chem. B* **2005**, *109*, 2805–2809. [CrossRef] [PubMed]
75. Yang, J.; Bai, H.; Tan, X.; Lian, J. IR and XPS investigation of visible-light photocatalysis-Nitrogen-carbon-doped TiO2 film. *Appl. Surf. Sci.* **2006**, *253*, 1988–1994. [CrossRef]
76. Chu, W.; Choy, W.K. The mechanisms of rate enhancing and quenching of trichloroethene photodecay in the presence of sensitizer and hydrogen sources. *Water Res.* **2002**, *36*, 2525–2532. [CrossRef]
77. Ollis, D.F.; Pelizzetti, E.; Serpone, N. Destruction of water contaminants. *Environ. Sci. Technol.* **1991**, *25*, 1522–1529. [CrossRef]
78. Ilisz, I.; Föglein, K.; Dombi, A. The photochemical behavior of hydrogen peroxide in near UV-irradiated aqueous TiO2 suspensions. *J. Mol. Catal. A Chem.* **1998**, *135*, 55–61. [CrossRef]
79. Dionysiou, D.D.; Suidan, M.T.; Bekou, E.; Baudin, I.; Lainé, J.M. Effect of ionic strength and hydrogen peroxide on the photocatalytic degradation of 4-chlorobenzoic acid in water. *Appl. Catal. B Environ.* **2000**, *26*, 153–171. [CrossRef]
80. Nagaveni, K.; Sivalingam, G.; Hegde, M.S.; Madras, G. Solar photocatalytic degradation of dyes: High activity of combustion synthesized nano TiO_2. *Appl. Catal. B Environ.* **2004**, *48*, 83–93. [CrossRef]
81. Toor, A.T.; Verma, A.; Jotshi, C.K.; Bajpai, P.K.; Singh, V. Photocatalytic degradation of Direct Yellow 12 dye using UV/TiO_2 in a shallow pond slurry reactor. *Dyes Pigments* **2006**, *68*, 53–60. [CrossRef]
82. Konstantinou, I.K.; Albanis, T.A. TiO_2-assisted photocatalytic degradation of azo dyes in aqueous solution: Kinetic and mechanistic investigations: A review. *Appl. Catal. B Environ.* **2004**, *49*, 1–14. [CrossRef]
83. Grzechulska, J.; Morawski, A.W. Photocatalytic decomposition of azo-dye acid black 1 in water over modified titanium dioxide. *Appl. Catal. B Environ.* **2002**, *36*, 45–51. [CrossRef]
84. Wang, X.; Wang, J.; Guo, P.; Guo, W.; Li, G. Chemical effect of swirling jet-induced cavitation: Degradation of rhodamine B in aqueous solution. *Ultrason. Sonochem.* **2008**, *15*, 357–363. [CrossRef] [PubMed]
85. Tang, S.K.; Teng, T.T.; Alkarkhi, A.F.M.; Li, Z. Sonocatalytic Degradation of Rhodamine B in Aqueous Solution in the Presence of TiO_2 Coated Activated Carbon. *APCBEE Procedia* **2012**, *1*, 110–115. [CrossRef]
86. Mcheik, H.A.; Jamal, M.M. Kinetic study of the discoloration of rhodamine B with persulfate, Iron activation. *J. Univ. Chem. Technol. Metall.* **2013**, *48*, 357–365.
87. Sap, K.A.; Demmers, J.A.A. World's largest Science, Technology & Medicine Open Access book publisher c. *Intech* **2016**, *6*, 111–133. [CrossRef]
88. Martha, S.; Sahoo, P.C.; Parida, K.M. An overview on visible light responsive metal oxide based photocatalysts for hydrogen energy production. *RSC Adv.* **2015**, *5*, 61535–61553. [CrossRef]
89. Tak, Y.; Kim, H.; Lee, D.; Yong, K. Type-II CdS nanoparticle–ZnO nanowire heterostructure arrays fabricated by a solution process: Enhanced photocatalytic activity. *Chem. Commun.* **2008**, 4585. [CrossRef]

90. Han, S.K.; Hwang, T.M.; Yoon, Y.; Kang, J.W. Evidence of singlet oxygen and hydroxyl radical formation in aqueous goethite suspension using spin-trapping electron paramagnetic resonance (EPR). *Chemosphere* **2011**, *84*, 1095–1101. [CrossRef]
91. Zhang, Y.; Barnes, A.N.; Zhu, X.; Campbell, N.F.; Gao, R. Quantification of thiopurine/UVA-induced singlet oxygen production. *J. Photochem. Photobiol. A Chem.* **2011**, *224*, 16–24. [CrossRef] [PubMed]
92. Jovanović, S.P.; Syrgiannis, Z.; Marković, Z.M.; Bonasera, A.; Kepić, D.P.; Budimir, M.D.; Milivojević, D.D.; Spasojević, V.D.; Dramićanin, M.D.; Pavlović, V.B.; et al. Modification of Structural and Luminescence Properties of Graphene Quantum Dots by Gamma Irradiation and Their Application in a Photodynamic Therapy. *ACS Appl. Mater. Interfaces* **2015**, *7*, 25865–25874. [CrossRef] [PubMed]
93. Li, L.; Liu, S.; Zhu, T. Application of activated carbon derived from scrap tires for adsorption of Rhodamine B. *J. Environ. Sci.* **2010**, *22*, 1273–1280. [CrossRef]
94. Gao, H.; Sun, Y.; Zhou, J.; Xu, R.; Duan, H. Mussel-inspired synthesis of polydopamine-functionalized graphene hydrogel as reusable adsorbents for water purification. *ACS Appl. Mater. Interfaces* **2013**, *5*, 425–432. [CrossRef] [PubMed]
95. Dutta, S.; Sahoo, R.; Ray, C.; Sarkar, S.; Jana, J.; Negishi, Y.; Pal, T. Biomolecule-mediated CdS-TiO_2-reduced graphene oxide ternary nanocomposites for efficient visible light-driven photocatalysis. *Dalton Trans.* **2014**, *44*, 193–201. [CrossRef] [PubMed]
96. Lv, T.; Pan, L.; Liu, X.; Lu, T.; Zhu, G.; Sun, Z.; Sun, C.Q. One-step synthesis of CdS–TiO_2–chemically reduced graphene oxide composites via microwave-assisted reaction for visible-light photocatalytic degradation of methyl orange. *Catal. Sci. Technol.* **2012**, *2*, 754. [CrossRef]
97. Egger, D.A.; Rappe, A.M.; Kronik, L. Hybrid Organic-Inorganic Perovskites on the Move. *Acc. Chem. Res.* **2016**, *49*, 573–581. [CrossRef]
98. Fu, H.; Pan, C.; Yao, W.; Zhu, Y. Visible-light-induced degradation of rhodamine B by nanosized Bi_2WO_6. *J. Phys. Chem. B* **2005**, *109*, 22432–22439. [CrossRef]

nanomaterials

MDPI

Article

Fabrication of ZnO/Red Phosphorus Heterostructure for Effective Photocatalytic H_2 Evolution from Water Splitting

Jiaqi Chen, Shaolong Huang, Yaojia Long, Jiahao Wu, Hui Li, Zhao Li, Yu-Jia Zeng * and Shuangchen Ruan

Shenzhen Key Laboratory of Laser Engineering, College of Optoelectronic Engineering, Shenzhen University, Shenzhen 518060, China; 2160190414@email.szu.edu.cn (J.C.); nkhsl3313@163.com (S.H.); longyj1110@126.com (Y.L.); wujh83@mail2.sysu.edu.cn (J.W.); hui75401@gmail.com (H.L.); lizhao_520@163.com (Z.L.); scruan@szu.edu.cn (S.R.)

* Correspondence: yjzeng@szu.edu.cn; Tel.: +86-136-1286-2990

Received: 17 September 2018; Accepted: 11 October 2018; Published: 15 October 2018

Abstract: Photocatalysis is a green technique that can convert solar energy to chemical energy, especially in H_2 production from water splitting. In this study, ZnO and red phosphorus (ZnO/RP) heterostructures were fabricated through a facile calcination method for the first time, which showed the considerable photocatalytic activity of H_2 evolution. The photocatalytic activities of heterostructures with different ratios of RP have been investigated in detail. Compared to bare ZnO, ZnO/RP heterostructures exhibit a 20.8-fold enhancement for H_2 production and furthermore overcome the photocorrosion issue of ZnO. The improved photocatalytic activities highly depend on the synergistic effect of the high migration efficiency of photo-induced electron–hole pairs with the inhibited charge carrier recombination on the surface. The presented strategy can also be applied to other semiconductors for various optoelectronics applications.

Keywords: photocatalysis; H_2 evolution; red P; ZnO; heterostructure

1. Introduction

To solve the current energy crisis, H_2 evolution from water splitting has been considered as one of the most promising methods for harvesting clean fuels. Since the pioneering work from Fujishima and Honda to induce the photo-assisted decomposition of water into H_2 by using UV light in 1972, the photocatalytic properties of semiconductors have been studied in detail to directly convert solar energy into solar fuels [1–3]. Metal oxide semiconductors, such as TiO_2 [4], WO_3 [5], ZrO_2 [6], SnO_2 [7], CeO_2 [8], ZnO [9], have been utilized as promising photo-catalyzers to generate H_2 [10]. Among them, ZnO is nontoxic, low-cost and eco-friendly, and has been investigated for photoactivity [11–13]. However, the photocatalytic efficiency of ZnO is hindered by several shortages [14,15], including the high recombination rate of charge carriers, fatal photocorrosion [16], and the limited absorption of the solar spectra.

It is believed that heterojunction photocatalysts can promote carrier separation, which results in improved photocatalytic properties [17–19]. To enhance the photocatalytic performance of ZnO, constructing a heterojunction with other materials is a general strategy [20,21]. Recently, many metal-free elemental photocatalysts (Si, Se, P, S, B, Te) have received particular attention due to their good photoactivity [22]. Among them, red phosphorus (RP) is considered to be a promising candidate as it is a cost-effective and earth-abundant element [23,24]. Recently, black P/red P heterojunctions have been synthesized by an in-situ mechanical milling method, which shows enhanced photocatalytic activity for RhB dye degradation [25]. Xue et al. prepared a RP/C_3N_4

heterojunction for photocatalytic H_2 production and CO_2 conversion [26]. Moreover, RP and CdS have been constructed as heterostructure photocatalysts with enhanced photocatalytic H_2 evolution activity [27]. However, heterostructure photocatalysts based on RP and ZnO have barely been studied.

In this study, ZnO/RP heterostructures have been prepared through a facile pressure-tight capsule calcination method, which shows enhanced photocatalytic H_2 evolution performance and good photostability. By optimizing the composition and microstructure, the as-prepared nanoparticles exhibit excellent photocatalytic stability thanks to the coating of RP particles. The significantly enhanced H_2 evolution rate is believed to result from the synergistic effect of the high migration efficiency of photo-induced electron–hole pairs with the inhibited charge carrier recombination in the interface.

2. Experiments

2.1. Synthesis of ZnO/RP Heterojunction Photocatalyst

Commercial ZnO (100 mg, Aladdin, Shanghai, China, AR 99.9%) was placed into SiO_2 capsules and put into capsules with 1 mg, 5 mg, 10 mg, 15 mg of commercial RP (Aladdin, Shanghai, China), respectively. By adjusting the used amount of RP, the coverage density of the RP on the surface of the ZnO nanoparticle was controlled to obtain different ratios, which were marked as ZRP-X (X = weight percentage of added RP). Followed by vacuuming and sealing, the capsules were heated to 550 °C for 4 h at a heating rate of 5 °C/min, and then cooled down to room temperature naturally. For comparison, the pure ZnO power was also treated with the same procedure in the absence of red P.

2.2. Characterizations

X-ray diffraction (XRD) patterns were measured on instrument (D8Advance, Bruker, Karlsruhe, Germany) using Cu-Kα-radiation. The acceleration voltage and the applied current were set as 40 kV and 40 mA, respectively. Scanning electron microscopy (SEM) was performed using a compact (SU70, Hitachi, Tokyo, Japan) with EDS mapping using Bruker XFlash 6I10 (XFlash 6I10, Bruker, Karlsruhe, Germany) at an accelerating voltage of 10 kV. Transmission electron microscopy (TEM) and high resolution TEM (HRTEM) analyses were measured on a JEM-2100&Aztec Energy TEM SP X-MaxN 80T (JEOL, Tokyo, Japan) using an accelerating voltage of 200 kV. The samples were prepared by applying a drop of ethanol suspension onto an amorphous carbon-coated copper grid and dried naturally. To figure out the surface chemical states, the X-ray photoelectron spectra (XPS) of the prepared photocatalysts were recorded on a Thermo Scientific ESCALAB 250Xi (ThermoFisher Scientific, Waltham, MA, USA). All spectra were calibrated to the binding energy of the adventitious C1s peak at 284.6 eV. UV-vis diffuse reflectance spectra (DRS) were recorded over the spectral range of 300–800 nm on a Perkin-Elmer Lambda 750 UV-vis spectrometer (Perkin Elmer, Waltham, MA, USA), using $BaSO_4$ as a reflectance standard. Specific surface area (SBET) was determined with a surface area analyzer (Nova 2000e, Quantachrome, FL, USA) from the Brunauer-Emmett-Teller (BET) theory. Photoluminescence (PL) and Raman spectra were obtained on a HORIBA Jobin Yvon LabRAM HR (Horiba, Kyoto, Japan) at an excitation wavelength of 325 nm and 514 nm, respectively. The transient fluorescence decay spectra were measured by Edinburgh Instruments FLS920 fluorescence spectrophotometer (Edinburgh Instruments, Edinburgh, UK) using the 325 nm line of the Xe lamp as the excitation source.

2.3. Photocatalytic H_2 Evolution Experiments

The photocatalytic H_2 evolution experiments were performed using a reaction (CLE-SPH2N, Aulight, Beijing, China) cell connected to a closed gas circulation and evacuation system. In a typical procedure, 50 mg of the sample was dispersed in 100 mL of deionized water, with 10 vol% Triethanolamine (TEOA, Aladdin, Shanghai, China) as the hole's sacrificial agent and 50 μL 3 wt% hydrochloroplatinic acid (ACS, 99.95%, Pt 37.5%, Alfa Aesar, Shanghai, China) as the co-catalyst.

After being exposed to ultrasonic conditions for 5 min to get a homogeneous solution, the water splitting experiment was measured in a closed gas recirculation system equipped with a quartz reactor, connected to an evacuation pump and irradiated by a 300 W xenon lamp (CEL-HXF300, Aulight, Beijing, China) under one-sun light by using an AM 1.5 solar filter to obtain a measured intensity equivalent to standard AM 1.5 sunlight (100 mW/cm^2). The light intensity was tested by an optical power meter (CEL-NP2000-10, Aulight, Beijing, China). The amount of H_2 generated from the photocatalytic reaction was measured by a Techcomp GC 7920 gas chromatograph equipped with a TCD detector (Techcomp GC 7920, Shanghai, China) every 30 min. High-purity nitrogen gas was used as the carrier gas. In the long-time stability test, the reaction system was tested every 6 h of irradiation without adding extra TEOA or Pt.

3. Results and Discussion

As illustrated in Figure 1a, the heterostructure photocatalysts were prepared by calcining the mixture of RP and ZnO in vacuum. With the increase of the content of RP, the color of the heterojunctions becomes darker (Figure S1). The morphology and crystal phase of the as-synthesized samples were investigated with SEM, as shown in Figure 1b–e. It can be observed from Figure 1b,c that ZRP-1 and ZRP-5 exhibit a hexagonal structure of ZnO, which is consistent with the previous report [28]. On the other hand, the morphologies of ZRP-10 and ZRP-15 change as shown in Figure 1d,e, suggesting the increase of RP concentration in the ZnO/RP heterostructures. It is widely accepted that the activity of a photocatalyst is closely related to its morphology and crystallinity. Therefore, the heterostructures with different RP concentrations as well as different morphologies are expected to exhibit different performances of H_2 evolution. The detailed structural information of ZnO/RP is shown in TEM images in Figure 2a. Figure 2b shows the HRTEM image of the heterostructure. The lattice spacing of 0.26 and 0.34 nm can be clearly observed, corresponding to the (101) plane of ZnO and (021) plane of RP, respectively [27]. These results confirm the formation of a heterostructure based on ZnO and RP. Furthermore, Figure 3a shows the Raman spectra of the ZnO, RP, and ZnO/RP heterostructures, where ZnO Raman modes are located at 350–550 cm^{-1} while the RP shows several well-defined modes in the 300–500 cm^{-1} region. For the ZnO/RP sample, a peak at 348 cm^{-1} related to RP is evident, indicateingthe existence of RP in the ZRP-5 heterostructure [29]. In Figure 3b, we show the XRD patterns of as-prepared samples. XRD patterns indicate that ZnO/RP heterostructures are well crystalline with a hexagonal structure, and the peaks centered at 31.7°, 34.5° and 47.5° can be indexed to the (100), (002) and (101) planes of hexagonal ZnO (PDF#80-0074). Due to the weak crystallization of RP, only one peak is found at 15° for RP corresponding to (102) planes (JCPDS card no. 44-0906) [27,29]. It can be considered that a proper amount of RP introduction does not change the phase structure of ZnO [30]. Compared to the pristine ZnO, no shift is detected in the ZnO/RP heterostructure, which confirms the formation of ZnO/RP heterojunction rather than RP doped ZnO.

The photocatalytic H_2 evolution rates of the samples (ZnO, ZRP-1, ZRP-5, ZRP-10 and ZRP-15) are measured under AM1.5 irradiation. As shown in Figure 4a, the H_2 evolution amounts of ZnO, ZRP-1, ZRP-5, ZRP-10 and ZRP-15 are evaluated to be 28.68, 442.57, 594.30, 391.33, and 261.33 μmol g^{-1} in the first 6 h. The H_2 evolution increases with the content of RP and reaches its optimal value when the proportion is 5 wt%. As shown in Figure 4b, ZRP-5 shows the highest photocatalytic activity in 6 h, which shows a 20.8-fold enhancement against the bare ZnO. Obviously, the amount of RP has a direct impact on the growth of the ZnO/RP [31]. Meanwhile, the low photoactivity of the pure ZnO can be attributed to its limited efficiency and the high recombination rate. The separation and transfer efficiency of the photo-generated charge carriers and recombination behavior of the photo-induced electron and hole are crucial to the photocatalytic H_2 evolution.

To investigate the separation and transfer efficiency of the photo-generated carriers over the ZnO/RP heterostructures, the PL spectra of the as-synthesized samples are shown in Figure 5a. It is known that the photo-induced electron in the conduction band will recombine with the hole at the valence band, leading to the emission of fluorescence [32]. Therefore, the quench of the PL spectra usually indicates the inhibition of the recombination for charge carriers, implying the separation of photo-induced electrons and holes. The pure ZnO shows two strong emission peaks: one is located at ~392 nm, corresponding to the near band gap excitonic emission, and the other is located at ~520 nm, attributed to the presence of singly ionized oxygen vacancies. In our ZnO/RP sample, both near band edge emission at 392 nm and defect-related emission at 520 nm decrease simultaneously. In addition, the photoactivity of the ZnO/RP sample increases as compared to the ZnO sample. Therefore, we believe that the decrease of PL is due to the formation of a heterostructure, which greatly increases the charge separation efficiency and reduces the recombination probability of photogenerated electron-holes [33,34]. Note that it has been reported that the defect-related emission at 475~625 nm disappears in the good-quality P-doped ZnO nanowires, which has been ascribed to the P doping effect instead of the formation of a heterojunction [35]. The results demonstrate an efficient separation of photo-excited electron–hole pairs between ZnO and RP, which is the main reason for enhancing the photocatalytic activity H_2 evolution by using the ZnO/RP heterostructure.

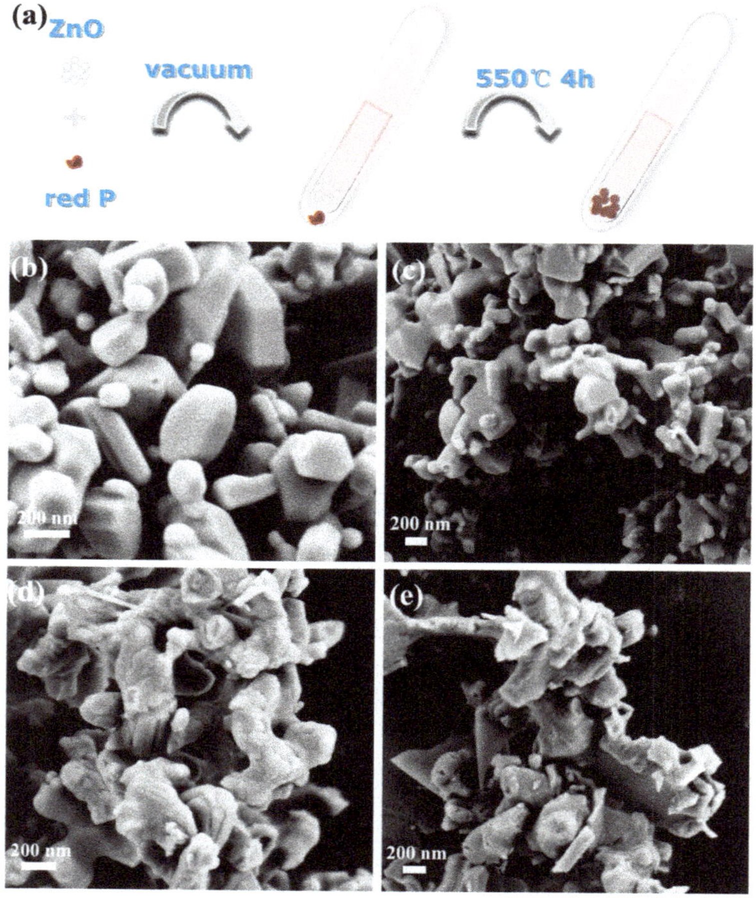

Figure 1. (**a**) Synthetic procedure for ZnO and red phosphorus (ZRP) heterostructure; (**b**–**e**) SEM images of ZnO-1, ZnO-5, ZnO-10 and ZnO-15.

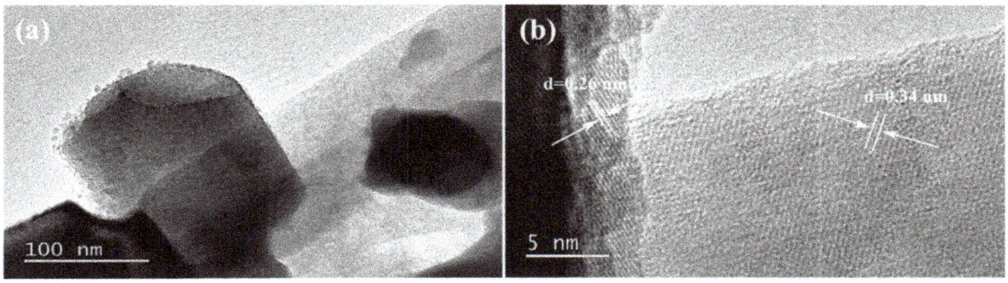

Figure 2. (**a**) TEM image of ZRP heterostructure; (**b**) HRTEM image of ZRP heterostructure.

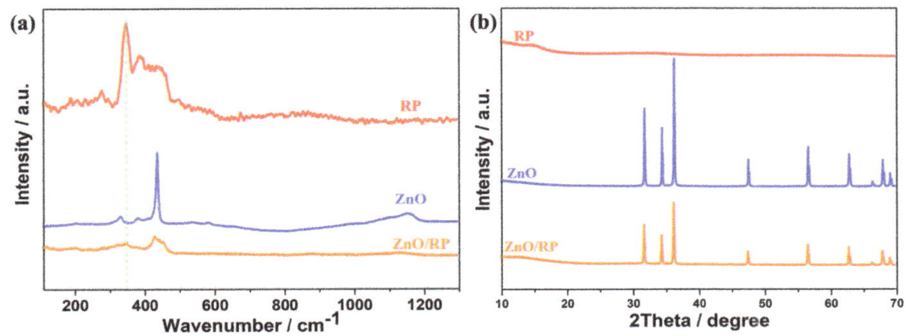

Figure 3. (**a**) Raman spectra of red P, ZnO and ZRP-5; (**b**) XRD patterns of the samples: ZnO, red P and ZRP-5.

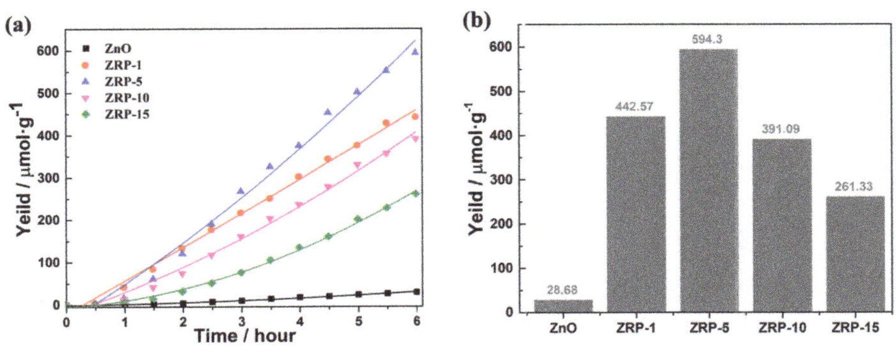

Figure 4. (**a**) Photocatalytic H_2 evolution of the as-synthesized heterojunctions under AM 1.5 irradiation; (**b**) Photocatalytic H_2 release yield in 6 h.

Transient fluorescence decay spectra have also been used to illustrate the recombination efficiency of photo-induced carriers. The lifetime and the percentage of the charge carrier are summarized in Figure 5b. The decay time with τ_1 (2.31 ns, 96%) and τ_2 (33.05 ns, 4%) of the transient fluorescence are detected in the pure ZnO with 325 nm excitation. On the other hand, in the ZnO/RP heterostructure, both τ_1 (2.81 ns, 78%) and τ_2 (43.54 ns 22%) increase as compared with those for pure ZnO. This is because a portion of electrons in the conduction band (CB) of ZnO recombine with the holes in the valence band (VB) of RP, resulting in a decreased recombination of photogenerated electron–hole pairs in ZnO and extending the lifetime of holes in the VB of ZnO [36,37]. As confirmed, the fast decay component τ_1 2.31 ns of ZnO/RP is mainly assigned to the decreased lifetime of the charge recombination in the heterojunction due to the improved charge redistribution on the heterointerface [38]. Moreover, the lifetime of τ_2 43.54 ns, which is due to the prolonged charge recombination process in the ZnO/RP can maintain the photocatalytic activity and reduce the recombination of electron–hole pairs. As a result, the formation of new interaction generates more effective charge separation and improves the photocatalytic activity of H_2 production.

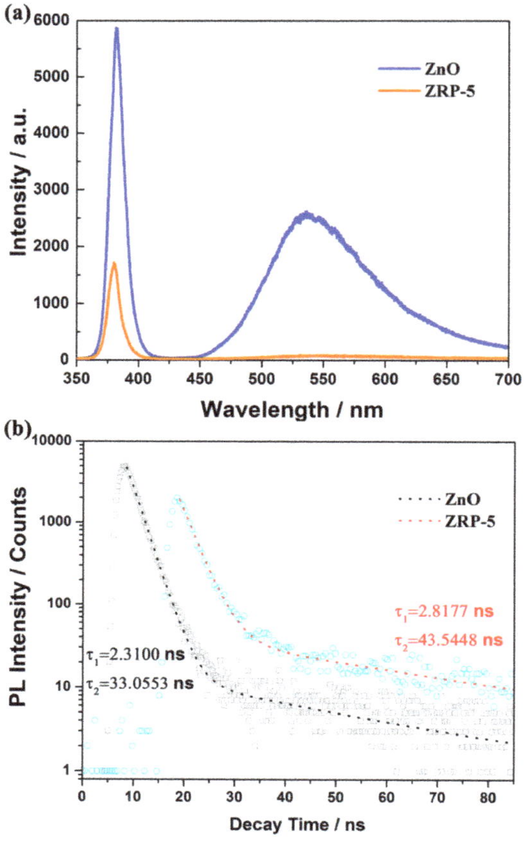

Figure 5. (**a**) Photoluminescence (PL) spectra of ZnO and ZRP-5; (**b**) Transient fluorescence decay spectra of ZnO and ZRP-5.

The photocatalytic activities of semiconductors are closely related to their structure and the inner electron behavior; thus, the UV−vis absorption spectra were measured. Figure 6a exhibits UV-vis absorption features of ZnO, RP and ZRP-5. Accordingly, RP and ZnO show a fundamental absorption edge at 719 nm and 373 nm, respectively. The high absorption efficiency of these heterojunction photocatalysts will harvest more photons into the photocatalytic reaction process to enhance the photocatalytic activity [39]. As expected, the absorption edge of the heterojunction ZRP-5 exhibits a subtle red shift to the visible light region. According to the Kubelka–Munk rule, the bandgap of RP and ZnO is calculated to be 1.54 and 3.27 eV [25,40], respectively. The presence of RP enhances the light absorption for the ZnO/RP sample significantly in the region of $\lambda > 400$ nm, which is due to the formation of ZnO/RP heterojunction and the interfacial interaction [27].

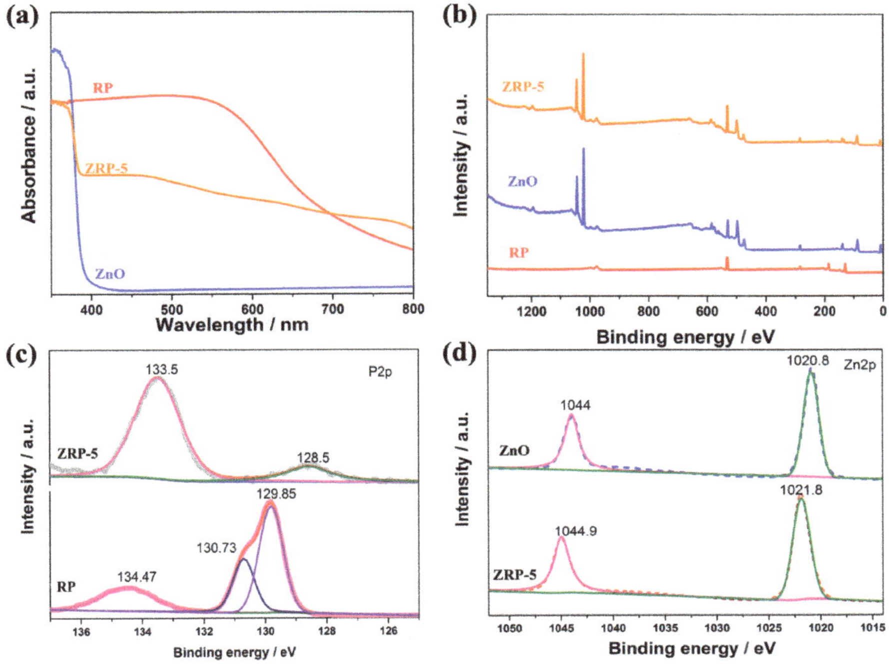

Figure 6. (**a**) UV-vis diffuse reflectance spectra of ZnO, red P and ZRP-5; (**b**) The XPS survey spectra of ZnO, red P and ZRP-5; (**c**) The XPS spectra P *2p* of ZRP-5 and red P; (**d**) The XPS Zn *2p* spectra of ZnO and ZRP-5.

XPS analysis was performed to investigate the chemical states of pure ZnO, ZRP-5 and RP, as shown in Figure 6b. The P 2*p* spectra of ZRP-5 and RP are shown in Figure 6c. For the RP, the spectrum shows that its surface is mainly composed of P^0 atoms (129.8 eV) and a certain amount of P^{5+} atoms (134.47 eV) [32,41]. For ZRP-5, there are two peaks of P 2*p* located at 133.5 and 128.5 eV, which is consistent with P^{5+} and P^0, respectively. Remarkably, the binding energy of P 2*p* shifts toward lower energies compared with that of the pristine RP, revealing that the charge transfer in these heterojunction photocatalysts. In Figure 6d, Zn 2*p* spectrum shows two strong peaks located at 1020.8 and 1044 eV for the pure ZnO. ZRP-5 it presents double peaks at 1021.8 and 1044.9 eV. These two peaks shift to the higher binding energies, indicating the strong interaction between ZnO and RP and the possible transfer of photogenerated charge carriers [32]. The XPS results reveal the strong interaction and chemical bonding between ZnO and RP, which are believed to result in the fast immigration of charge carriers and the separation of photo-excited electron–hole pairs.

To investigate the effect on the surface area and pore structure of ZnO/RP, the N_2 adsorption–desorption isotherms were measured, and the textural parameters derived from the isotherms data are summarized in Table S1. The BET specific surface area of pure ZnO and ZRP-5 are 11.85 m^2/g and 16.24 m^2/g, respectively. However, we cannot find a direct correlation between photocatalytic H_2 evolution activity between the BET surface area and pore volume, which excludes the possibility that the BET surface area and pore volume are the crucial factors for the improved photocatalytic activity [32].

For comparison, the photoactivity of the mechanical mixture of ZnO and RP was also investigated. Figure S2 displays the photocatalytic H_2 production of the mechanical mixture sample, which shows negligible H_2 evolution, even smaller that of pure ZnO, probably due to the shield effect [41]. The results highlight the important role the interface of the ZnO/RP heterostructure, which is crucial to the formation of electronic interaction and electron transfer. Note that no photocatalytic activity has been observed for the pure commercial red phosphorus used in this study. ZRP-X heterostructures show obviously increased H_2 evolution rates due to the improved light absorption and the low recombination of photogenerated carriers resulting from the heterojunction structure [13]. However, with the increasing concentration of RP (>5 wt %) in the heterostructure, the photocatalytic reactivity appears to decrease. This might be attributed to the serious agglomeration of RP particles, leading to the blocking of the light absorption [27].

As a typical photocatalyst, the instability has been a critical issue for ZnO because of the photocorrosion [23]. Figure S3a shows the evolution curves of H_2 in a cycling photocatalytic three times to test the stability of the heterojunction. The ZnO/RP heterostructure tends to be stable, and no quick decrease is observed after the test, showing the good photostability of the heterojunction due to the introduction of RP. In addition, the XPS patterns Figure S3b of ZRP-5 before and after 18 h irradiation are determined to investigate the structure stability, and no distinguishable change can be observed, providing evidence for the improved stability of heterojunctions.

In order to determine the photocatalytic mechanism of the heterostructure, the ZRP-5 sample is measured by valence band XPS, and the valance band minimums (VBMs) of the samples are determined in Figure S4. Both ZnO and ZRP-5 samples display a typical VBM at about 2.27 eV. However, compared with the pure ZnO, a small peak within the range of 0.5–2.0 eV is located at 1.2 eV for ZRP-5, which indicates that the VBM of RP is approximately located at 1.07 eV. Based on their bandgap values, the band alignment of the heterojunction belongs to type-I.

The type-I heterojunction in Figure 7 is considered to be the possible mechanism to explain the improved photocatalytic activity [20]. For the type-I configuration in our work, under the light irradiation, electrons and holes will accumulate in the conduction band minimum (CBM) and VBM of RP, respectively. Therefore, efficient band alignment can be used to separate the charge carriers and reduce the carrier recombination, which results in significant enhancement of photocatalytic activity. Besides this, chloroplatinic acid was added before the photocatalytic reaction. Under AM 1.5 radiation, the Pt^+ ions are reduced to Pt nanoparticles at the reduction sites. Therefore, the reaction site can be determined by the Pt distribution on the samples after the reaction. After the photocatalyst reaction, Pt is loaded on the RP as confirmed by the EDS mapping (Figure S5): this result indicates that the reduction reaction occurs on the RP site, which is consistent with the mechanism discussed above.

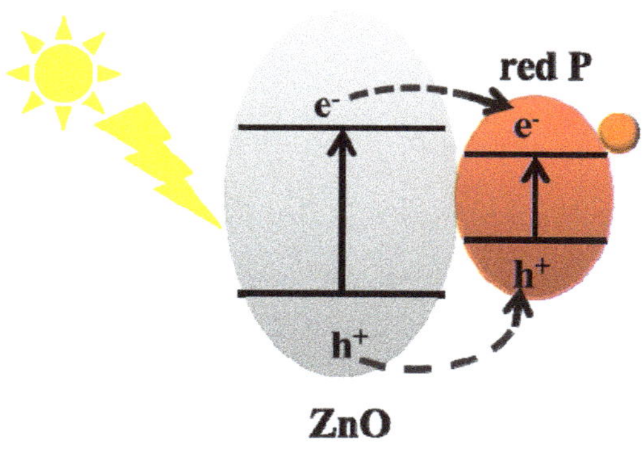

Figure 7. Type—band alignment of ZnO and RP.

4. Conclusions

We have developed a novel kind of type-I ZnO/RP heterostructure photocatalyst through a simple calcination method under vacuum. Compared to the pure ZnO and red P, ZnO/RP heterostructures exhibit enhanced photocatalytic activity for H_2 evolution and excellent photostability under AM1.5 light irradiation. As confirmed by PL and transient fluorescence spectra, the enhancement of water splitting to H_2 evolution is believed to result from the rapid transfer and effective separation of photogenerated electrons and holes between the heterointerface of ZnO and RP. Thus, the increased charge carrier lifetime and the decreased recombination rate of the photogenerated electron–hole pairs both contribute to the enhancement of photocatalytic activity. This work not only demonstrates a photocatalyst based on ZnO/RP heterostructure, but also provides a simple strategy to construct heterojunctions that can also be applied to other semiconductors for optoelectronics applications.

Supplementary Materials: The following are available online at http://www.mdpi.com/2079-4991/8/10/835/s1, Figure S1: Digital photograph of the heterostructure (A:ZRP-1, B:ZRP-5, C:ZRP-10, D:ZRP-15), Figure S2: The photocatalytic H2 production of red P, ZnO, and mechanical mixture (the content of red P is 5%), Figure S3: (a) The recycling H2 evolution reaction of ZRP-5 heterojunction. (b)The compared XPS pattern of ZRP-5 before and after irradiation, Figure S4: The valence band XPS of ZnO and ZRP-5, Figure S5: (a) SEM image of recycled ZRP-5 sample after water splitting photocatalytic reaction. (b–d) EDS mapping of O, Zn, P and Pt. (f) EDS spectra; Table S1: Specific surface area and pore volume of the samples.

Author Contributions: Conceptualization, J.C.; Methodology, J.C..; Software, J.W..; Validation, S.H. and Y.L; Formal Analysis, J.C.; Investigation, J.C..; Resources, S.R..; Data Curation, J.C. and H.L.; Writing-Original Draft Preparation, J.C.; Writing-Review & Editing, S.H. and Y.-J.Z.; Visualization, Z.L.; Supervision, Y.-J.Z. and S.R..; Project Administration, Y.-J.Z.; Funding Acquisition, Y.-J.Z. and S.R.

Acknowledgments: This work was supported by the National Natural Science Foundation of China under Grant No. 51502178, the Major Science and Technology Project of Guangdong Province No. 2014B010131006, the Shenzhen Science and Technology Project under Grant Nos. JCYJ20170412105400428 and KQJSCX20170727101208249.

Conflicts of Interest: The authors declare no conflict of interest.

References

1. Yu, C.; Yu, J.C.; Zhou, W.; Yang, K. WO_3 Coupled P-TiO_2 photocatalysts with mesoporous structure. *Catal. Lett.* **2010**, *140*, 172–183. [CrossRef]
2. Ran, J.; Zhang, J.; Yu, J.; Jaroniec, M.; Qiao, S.Z. Earth-abundant cocatalysts for semiconductor-based photocatalytic water splitting. *Chem. Soc. Rev.* **2014**, *43*, 7787–7812. [CrossRef] [PubMed]
3. Fujishima, A.; Honda, K. Electrochemical photolysis of water at a semiconductor electrode. *Nature* **1972**, *238*, 238–245. [CrossRef]
4. Yang, Y.; Liu, G.; Irvine, J.T.; Cheng, H.M. Enhanced photocatalytic H_2 production in Core-Shell engineered rutile TiO_2. *Adv. Mater.* **2016**, *28*, 5850–5856. [CrossRef] [PubMed]
5. Jungwon, K.; Chulwee, L.; Wonyong, C. Platinized WO_3 as an environmental photocatalyst that generates OH radicals under visible light. *Environ. Sci. Technol.* **2010**, *44*, 6849. [CrossRef]
6. Kambur, A.; Pozan, G.S.; Boz, I. Preparation, characterization and photocatalytic activity of TiO_2–ZrO_2 binary oxide nanoparticles. *Appl. Catal. B* **2012**, *115–116*, 149–158. [CrossRef]
7. Dodd, A.; McKinley, A.; Saunders, M.; Tsuzuki, T. Mechanochemical synthesis of nanocrystalline SnO_2–ZnO photocatalysts. *Nanotechnology* **2006**, *17*, 692–698. [CrossRef]
8. Zhang, N.; Fu, X.; Xu, Y.-J. A facile and green approach to synthesize Pt@CeO_2 nanocomposite with tunable core-shell and yolk-shell structure and its application as a visible light photocatalyst. *J. Mater. Chem.* **2011**, *21*, 8152. [CrossRef]
9. Li, Y.; Wang, L.; Liang, J.; Gao, F.; Yin, K.; Dai, P. Hierarchical heterostructure of ZnO@TiO_2 hollow spheres for highly efficient photocatalytic hydrogen evolution. *Nanoscale Res. Lett.* **2017**, *12*, 531. [CrossRef] [PubMed]
10. Khan, M.M.; Adil, S.F.; Al-Mayouf, A. Metal oxides as photocatalysts. *J. Saudi Chem. Soc.* **2015**, *19*, 462–464. [CrossRef]
11. Lee, K.M.; Lai, C.W.; Ngai, K.S.; Juan, J.C. Recent developments of zinc oxide based photocatalyst in water treatment technology: A review. *Water Res.* **2016**, *88*, 428–448. [CrossRef] [PubMed]

12. Zeng, Y.J.; Pereira, L.M.; Menghini, M.; Temst, K.; Vantomme, A.; Locquet, J.P.; Van Haesendonck, C. Tuning quantum corrections and magnetoresistance in ZnO nanowires by ion implantation. *Nano Lett.* **2012**, *12*, 666–672. [CrossRef] [PubMed]
13. Zeng, Y.J.; Ye, Z.Z.; Lu, Y.F.; Lu, J.G.; Sun, L.; Xu, W.Z.; Zhu, L.P.; Zhao, B.H.; Che, Y. ZnMgO quantum dots grown by low-pressure metal organic chemical vapor deposition. *Appl. Phys. Lett.* **2007**, *90*, 012111. [CrossRef]
14. Ma, D.; Shi, J.W.; Zou, Y.; Fan, Z.; Ji, X.; Niu, C. Highly efficient photocatalyst based on a CdS Quantum Dots/ZnO nanosheets 0D/2D heterojunction for hydrogen evolution from water splitting. *ACS. Appl. Mater. Inter.* **2017**, *9*, 25377–25386. [CrossRef] [PubMed]
15. Wang, X.; Liu, G.; Chen, Z.G.; Li, F.; Wang, L.; Lu, G.Q.; Cheng, H.M. Enhanced photocatalytic hydrogen evolution by prolonging the lifetime of carriers in ZnO/CdS heterostructures. *Chem. Commun.* **2009**, *23*, 3452–3454. [CrossRef] [PubMed]
16. Zhang, L.; Cheng, H.; Zong, R.; Zhu, Y. Photocorrosion Suppression of ZnO Nanoparticles via Hybridization with Graphite-like Carbon and Enhanced Photocatalytic Activity. *J. Phys. Chem. C* **2009**, *113*, 2368–2374. [CrossRef]
17. Wang, H.; Zhang, L.; Chen, Z.; Hu, J.; Li, S.; Wang, Z.; Liu, J.; Wang, X. Semiconductor heterojunction photocatalysts: Design, construction, and photocatalytic performances. *Chem. Soc. Rev.* **2014**, *43*, 5234–5244. [CrossRef] [PubMed]
18. Fujishima, A.; Zhang, X.; Tryk, D. Heterogeneous photocatalysis: From water photolysis to applications in environmental clean up. *Int. J. Hydrogen Energy* **2007**, *32*, 2664–2672. [CrossRef]
19. Low, J.; Cao, S.; Yu, J.; Wageh, S. Two-dimensional layered composite photocatalysts. *Chem. Commun.* **2014**, *50*, 10768–10777. [CrossRef] [PubMed]
20. Low, J.; Yu, J.; Jaroniec, M.; Wageh, S.; Al-Ghamdi, A.A. Heterojunction Photocatalysts. *Adv. Mater.* **2017**, *29*, 101694. [CrossRef] [PubMed]
21. Wang, Y.; Wang, Q.; Zhan, X.; Wang, F.; Safdar, M.; He, J. Visible light driven type II heterostructures and their enhanced photocatalysis properties: A review. *Nanoscale* **2013**, *5*, 8326–8339. [CrossRef] [PubMed]
22. Hu, Z.; Shen, Z.; Yu, J.C. Phosphorus containing materials for photocatalytic hydrogen evolution. *Green Chem.* **2017**, *19*, 588–613. [CrossRef]
23. Hu, Z.; Yuan, L.; Liu, Z.; Shen, Z.; Yu, J.C. An elemental phosphorus photocatalyst with a record high hydrogen evolution efficiency. *Angew. Chem. Int. Ed.* **2016**, *55*, 9580–9585. [CrossRef] [PubMed]
24. Wang, F.; Ng, W.K.H.; Yu, J.C.; Zhu, H.; Li, C.; Zhang, L.; Liu, Z.; Li, Q. Red phosphorus: An elemental photocatalyst for hydrogen formation from water. *Appl. Catal. B* **2012**, *111–112*, 409–414. [CrossRef]
25. Shen, Z.; Sun, S.; Wang, W.; Liu, J.; Liu, Z.; Yu, J.C. A black–red phosphorus heterostructure for efficient visible-light-driven photocatalysis. *J. Mater. Chem. A* **2015**, *3*, 3285–3288. [CrossRef]
26. Yuan, Y.-P.; Cao, S.-W.; Liao, Y.-S.; Yin, L.-S.; Xue, C. Red phosphor/g-C_3N_4 heterojunction with enhanced photocatalytic activities for solar fuels production. *Appl. Catal. B* **2013**, *140–141*, 164–168. [CrossRef]
27. Shi, Z.; Dong, X.; Dang, H. Facile fabrication of novel red phosphorus-CdS composite photocatalysts for H_2 evolution under visible light irradiation. *Int. J. Hydrogen Energy* **2016**, *41*, 5908–5915. [CrossRef]
28. Qi, K.; Yang, J.; Fu, J.; Wang, G.; Zhu, L.; Liu, G.; Zheng, W. Morphology-controllable ZnO rings: Ionic liquid-assisted hydrothermal synthesis, growth mechanism and photoluminescence properties. *CrystEngComm* **2013**, *15*, 6729. [CrossRef]
29. Chang, W.C.; Tseng, K.W.; Tuan, H.Y. Solution Synthesis of Iodine-Doped Red Phosphorus Nanoparticles for Lithium-Ion Battery Anodes. *Nano Lett.* **2017**, *17*, 1240–1247. [CrossRef] [PubMed]
30. D'Amato, C.A.; Giovannetti, R.; Zannotti, M.; Rommozzi, E.; Minicucci, M.; Gunnella, R.; Di Cicco, A. Band gap implications on nano-TiO_2 surface modification with ascorbic acid for visible light-active polypropylene coated photocatalyst. *Nanomaterials* **2018**, *8*, 599. [CrossRef] [PubMed]
31. Gong, C.; Du, J.; Li, X.; Yu, Z.; Ma, J.; Qi, W.; Zhang, K.; Yang, J.; Luo, M.; Peng, H. One-step acidic hydrothermal preparation of dendritic rutile TiO_2 nanorods for Photocatalytic Performance. *Nanomaterials* **2018**, *8*, 683. [CrossRef] [PubMed]
32. Ma, D.; Shi, J.-W.; Zou, Y.; Fan, Z.; Ji, X.; Niu, C.; Wang, L. Rational design of CdS@ZnO core-shell structure via atomic layer deposition for drastically enhanced photocatalytic H_2 evolution with excellent photostability. *Nano Energy* **2017**, *39*, 183–191. [CrossRef]

33. Yu, Z.B.; Xie, Y.P.; Liu, G.; Lu, G.Q.; Ma, X.L.; Cheng, H.-M. Self-assembled CdS/Au/ZnO heterostructure induced by surface polar charges for efficient photocatalytic hydrogen evolution. *J. Mater. Chem. A* **2013**, *1*, 2773. [CrossRef]
34. Moon, S.C.; Mametsuka, H.; Tabata, S.; Suzuki, E. Photocatalytic production of hydrogen from water using TiO_2 and B/TiO_2. *Catal. Today* **2000**, *58*, 125–132. [CrossRef]
35. Gao, J.; Zhao, Q.; Sun, Y.; Li, G.; Zhang, J.; Yu, D. A Novel Way for Synthesizing Phosphorus-Doped Zno Nanowires. *Nanoscale Res. Lett.* **2011**, *6*, 45. [CrossRef] [PubMed]
36. Liu, C.; Zhang, Y.; Dong, F.; Reshak, A.H.; Ye, L.; Pinna, N.; Zeng, C.; Zhang, T.; Huang, H. Chlorine intercalation in graphitic carbon nitride for efficient photocatalysis. *Appl. Catal. B* **2017**, *203*, 465–474. [CrossRef]
37. Xu, T.; Zhang, L.; Cheng, H.; Zhu, Y. Significantly enhanced photocatalytic performance of ZnO via graphene hybridization and the mechanism study. *Appl. Catal. B* **2011**, *101*, 382–387. [CrossRef]
38. Yang, C.; Qin, J.; Xue, Z.; Ma, M.; Zhang, X.; Liu, R. Rational design of carbon-doped TiO_2 modified g-C_3N_4 via in-situ heat treatment for drastically improved photocatalytic hydrogen with excellent photostability. *Nano Energy* **2017**, *41*, 1–9. [CrossRef]
39. Zhang, Z.; Huang, J.; Fang, Y.; Zhang, M.; Liu, K.; Dong, B. A nonmetal plasmonic Z-Scheme photocatalyst with UV-to NIR-Driven photocatalytic protons reduction. *Adv. Mater.* **2017**, *29*. [CrossRef] [PubMed]
40. Gao, J.; Zhang, X.; Sun, Y.; Zhao, Q.; Yu, D. Compensation mechanism in N-doped ZnO nanowires. *Nanotechnology* **2010**, *21*, 245703. [CrossRef] [PubMed]
41. Jing, L.; Zhu, R.; Phillips, D.L.; Yu, J.C. Effective prevention of charge trapping in graphitic carbon nitride with nanosized red phosphorus modification for superior photo(electro)catalysis. *Adv. Funct. Mater.* **2017**, *27*, 1703484. [CrossRef]

nanomaterials

MDPI

Article

MXene Boosted CoNi-ZIF-67 as Highly Efficient Electrocatalysts for Oxygen Evolution

Yangyang Wen *, Zhiting Wei, Chang Ma, Xiaofei Xing, Zhenxing Li * and Dan Luo *

State Key Laboratory of Heavy Oil Processing, College of New Energy and Material, Beijing Key Laboratory of Biogas Upgrading Utilization, China University of Petroleum (Beijing), Beijing 102249, China; WZT1215385749@126.com (Z.W.); machang_cup@126.com (C.M.); xingxiaofei_cup@126.com (X.X.)

* Correspondence: wenyangyang@cup.edu.cn (Y.W.); lizx@cup.edu.cn (Z.L.); luodan@iccas.ac.cn (D.L.)

Received: 19 April 2019; Accepted: 15 May 2019; Published: 20 May 2019

Abstract: Oxygen evolution reaction (OER) is a pivotal step for many sustainable energy technologies, and exploring inexpensive and highly efficient electrocatalysts is one of the most crucial but challenging issues to overcome the sluggish kinetics and high overpotentials during OER. Among the numerous electrocatalysts, metal-organic frameworks (MOFs) have emerged as promising due to their high specific surface area, tunable porosity, and diversity of metal centers and functional groups. It is believed that combining MOFs with conductive nanostructures could significantly improve their catalytic activities. In this study, an MXene supported CoNi-ZIF-67 hybrid (CoNi-ZIF-67@$Ti_3C_2T_x$) was synthesized through the in-situ growth of bimetallic CoNi-ZIF-67 rhombic dodecahedrons on the $Ti_3C_2T_x$ matrix via a coprecipitation reaction. It is revealed that the inclusion of the MXene matrix not only produces smaller CoNi-ZIF-67 particles, but also increases the average oxidation of Co/Ni elements, endowing the CoNi-ZIF-67@$Ti_3C_2T_x$ as an excellent OER electrocatalyst. The effective synergy of the electrochemically active CoNi-ZIF-67 phase and highly conductive MXene support prompts the hybrid to process a superior OER catalytic activity with a low onset potential (275 mV vs. a reversible hydrogen electrode, RHE) and Tafel slope (65.1 mV·dec^{-1}), much better than the IrO_2 catalysts and the pure CoNi-ZIF-67. This work may pave a new way for developing efficient non-precious metal catalyst materials.

Keywords: oxygen evolution reaction; metal-organic frameworks; MXene; $Ti_3C_2T_x$; hybrid

1. Introduction

With the rapid combustion of fossil fuels and the ever-growing concerns relating to the environmental crisis, developing sustainable energy technologies (such as metal-air batteries and water splitting) has triggered extensive attention [1]. Oxygen evolution reaction (OER) is the key process for these electricity-driven devices, but it has been significantly hindered by its sluggish kinetics and substantial overpotential [2]. Therefore, highly active electrocatalysts are required to increase the reaction rate and to lower the overpotentials in the OER process. To date, the precious metal oxides (e.g., RuO_2 and IrO_2) are the best electrocatalysts with a promoted proton-coupled charge transfer process, but their scale-up implementation has been greatly hampered by their high price, scarcity and poor durability [3,4]. Within this context, increasing efforts have been devoted to the exploration of inexpensive, earth-abundant and highly efficient electrocatalysts for OER [5]. Among them, the earth-based transition metal-rich compounds, including transition metal oxides [3], sulfides [6] and phosphides [7], have exhibited great promise as OER electrocatalysts.

Recently, metal organic frameworks (MOFs) consisting of the coordination of organic ligands and metal ions or clusters have received increasing attention for catalysis-related applications [4,8]. MOFs are an important class of porous solids in electrocatalysis in view of their high specific surface area, tunable porosity, and diversity of metal centers and functional groups [9]. However, it is still a challenge to directly utilize MOFs as efficient OER electrocatalysts because of their poor conductivity. One commonly-used strategy is adopting MOFs as precursors to prepare metal-based compounds/porous carbon composites via a high-temperature pyrolysis. However, the active sites and intrinsic structure of MOFs are inevitably sacrificed with the loss of organic ligands during the pyrolysis [10]. Another possible strategy is to combine MOFs with conductive nanostructures, which has demonstrated a significant enhancement in the electrocatalytic properties [11,12].

MXene is a new class of two-dimensional materials, synthesized by selectively etching A layers from its MAX phase [13]. It can be represented using a formula of $M_{n+1}X_nT_x$, where M is the early transition metal, X stands for C and/or N elements, and T for the surface terminations (–O, –F or –OH) [14]. MXene has emerged as a promising nanomaterial in various fields, including energy storage [15,16], energy conversion [11,17], water purification [18], electromagnetic interference [19], and so on, owing to its excellent electrical conductivity and surface hydrophilicity. Besides, MXene also demonstrated the possibility of being an excellent support by altering the electrophilicity of active centers in the supported catalysts and thus modifying the catalytic activity of the composites [20].

In this work, a MXene supported CoNi-ZIF-67 hybrid (CoNi-ZIF-67@$Ti_3C_2T_x$) was synthesized by the in-situ growth of bimetallic CoNi-ZIF-67 rhombic dodecahedrons on the $Ti_3C_2T_x$ matrix via a coprecipitation reaction. The effective synergy of the CoNi-ZIF-67 and MXene phases endows the hybrid with a remarkable electrocatalytic activity for OER, with a low onset potential (275 mV vs. a reversible hydrogen electrode, RHE) and Tafel slope (65.1 mV·dec^{-1}).

2. Materials and Methods

2.1. Preparation of $Ti_3C_2T_x$ MXene

Ti_3AlC_2 powder was first prepared via the HF-etching method [21]. Briefly, 1 g of Ti_3AlC_2 powder was blended with 40 mL of 40 wt.% HF solution under continuous stirring at 45 °C for 24 h. The resulting suspension was separated by centrifugation, washed several times with distilled water, and freeze-dried, obtaining the accordion-like $Ti_3C_2T_x$ MXene.

2.2. Preparation of CoNi-ZIF-67@$Ti_3C_2T_x$ and Pure CoNi-ZIF-67

Typically, 300 mg $Ti_3C_2T_x$, 0.9 mmol $Co(NO_3)_2·6H_2O$ and 0.1 mmol $Ni(NO_3)_2·6H_2O$ were dispersed in 8 mL methanol under sonication for 1 h. Then, 8 mmol of 2-methylimidazole was dissolved in another 8 mL methanol under stirring for 30 min. The two above solutions were mixed together, followed by adding 2 mg of hexadecyl trimethyl ammonium bromide (CTAB) and continuously stirring for 8 h at room temperature. The final precipitates were collected by centrifugation, washed with methanol and water several times, and dried at 60 °C under vacuum for 12 h. As a control, pure CoNi-ZIF-67 was prepared via the same procedure but without adding the $Ti_3C_2T_x$.

2.3. Materials Characterizations

The morphology and structure of the as-prepared catalysts were characterized by scanning electron microscopy (SEM, Hitachi SU8010, Tokyo, Japan), transmission electron microscopy (TEM, JEM 2100 LaB6, Tokyo, Japan), powder X-ray diffractometer analysis (XRD, Bruker D8 Advance instrument, Karlsruhe, Germany) with a Cu Kα irradiation source at a scanning rate of 1° per min, and X-ray photoelectron spectroscopy (XPS, PHI5000 Versaprobe, Kanagawa, Japan) with an Al Kα X-ray source. The binding energies of the XPS measurements were calibrated to the C 1s peak at 285.0 eV. The specific surface areas and pore size distribution of the catalysts were conducted on the ASAP2460 Surface Area and Porosity Analyzer (Micromeritics, Atlanta, GA, USA). The surface areas (S_{BET}) were calculated from the N_2 sorption isotherms via the Brunauer-Emmett-Teller method, and the pore size distributions were calculated from the N_2 isotherms using the non-local density functional theory (NLDFT) method.

2.4. Electrode Preparation and Electrochemical Measurements

All electrocatalytic performances were evaluated on a CHI 760E electrochemical workstation (Chenhua Instrument, Shanghai, China) with a standard three-electrode system in 0.1 M KOH aqueous solution at room temperature. A glassy carbon electrode (GCE, 5 mm in diameter) coated with the as-prepared catalysts was employed as the working electrode, a Hg/HgO electrode as the reference electrode and a graphite rod as the counter electrode. Before the test, the catalyst ink was prepared by dispersing 10 mg of catalyst powder in a mixture of 40 μL 5 wt.% Nafion solution (Sigma-Aldrich, Shanghai, China), 750 μL water and 250 μL ethanol. After ultrasonication for 30 min, 10 μL of the catalyst ink was pipetted onto the freshly-polished GCE with a catalyst mass loading of ~0.5 $mg_{cat}\cdot cm^{-2}$. All the potentials were calibrated to a reversible hydrogen electrode (RHE) according to the equation, $E(\text{RHE}) = E(\text{Hg/HgO}) + 0.059\text{pH} + 0.098$. Before the electrochemical measurement, the electrolyte was bubbled with an O_2 flow for 30 min, and a gas flow was maintained over the electrolyte during the measurement to ensure the O_2 saturation. The polarization curves were tested using the linear sweep voltammetry (LSV) at a scan rate of 50 $mV\cdot s^{-1}$. The double-layer capacitance (C_{dl}) was calculated from the cyclic voltammetry (CV) curves in a small potential range of 1.023–1.073 V vs. RHE without the occurrence of an apparent faradic process. The plots of the current density difference [$\Delta J = (J\text{a} - J\text{c})$], at 1.048 V vs. RHE against the scan rates of 10–60 $mV\cdot s^{-1}$, were linearly fitted, and the slope is the C_{dl} of the catalysts. Electrochemical impedance spectroscopy (EIS) was carried out at 1.46 V vs. RHE in a frequency range of 0.1–10^5 Hz. For the stability test, the catalysts were performed at 1.46 V vs. RHE over a 20,000 s continuous time. In comparison, the commercial IrO_2 catalyst purchased from Sigma-Aldrich with the same catalyst mass loading was tested under the same conditions.

3. Results

The preparation procedure of CoNi-ZIF-67@$Ti_3C_2T_x$ is illustrated in Figure 1. In brief, the accordion-like $Ti_3C_2T_x$ was first prepared by the selective etching of Al layers from the Ti_3AlC_2 MAX phase using HF. Then, Co^{2+} and Ni^{2+} ions with 2-methylimidazole in methanol were added. The bimetallic CoNi-ZIF-67 could grow in-situ on the $Ti_3C_2T_x$ via a coprecipitation reaction. Considering the negatively charged $Ti_3C_2T_x$ surface due to the presence of numerous surface termination groups (e.g., –O, –OH, and –F) introduced during the etching process, Co^{2+} and Ni^{2+} ions could be easily adsorbed on these termination group sites by electrostatic interaction, and could in-situ synthesize CoNi-ZIF-67 rhombic dodecahedrons on the surface and between the interlayers of $Ti_3C_2T_x$.

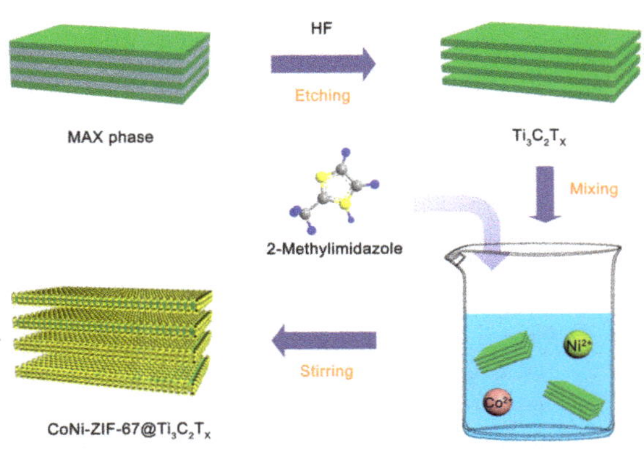

Figure 1. Schematic illustration of the preparation of CoNi-ZIF-67@$Ti_3C_2T_x$.

The XRD analysis was investigated for the structural characterization of the pristine $Ti_3C_2T_x$, CoNi-ZIF-67@$Ti_3C_2T_x$, and pure CoNi-ZIF-67. As shown in Figure 2a (enlarged image in Figure S1), the XRD pattern of the pristine $Ti_3C_2T_x$ represents the characteristic strong peak of the (002) plane at 8.2°, and the weak peaks of the (004), (101) and (110) planes according to the JCPDS card no. 52-0875, confirming the successful preparation of the $Ti_3C_2T_x$ phases. The pure CoNi-ZIF-67 shows the typical sharp peaks, consistent with the reported literatures [22]. The XRD pattern of the CoNi-ZIF-67@$Ti_3C_2T_x$ hybrid displays a superimposition of the two phases, featured with four obvious peaks at 6.6° of the (002) plane and 61.2° of the (110) plane for $Ti_3C_2T_x$, and 7.4° of the (011) plane and 12.8° of the (112) plane for CoNi-ZIF-67, revealing the effective combination of the CoNi-ZIF-67 and $Ti_3C_2T_x$ phases. It notes that an apparent shift of the (002) plane to a lower angle was detected in the CoNi-ZIF-67@$Ti_3C_2T_x$, compared with the pristine $Ti_3C_2T_x$ phase. This left-shift suggests a c-lattice parameter change from 2.16 nm in $Ti_3C_2T_x$ to 2.68 nm in the hybrid, disclosing the intercalation of the $Ti_3C_2T_x$ layers due to the inclusion of the CoNi-ZIF-67 particles.

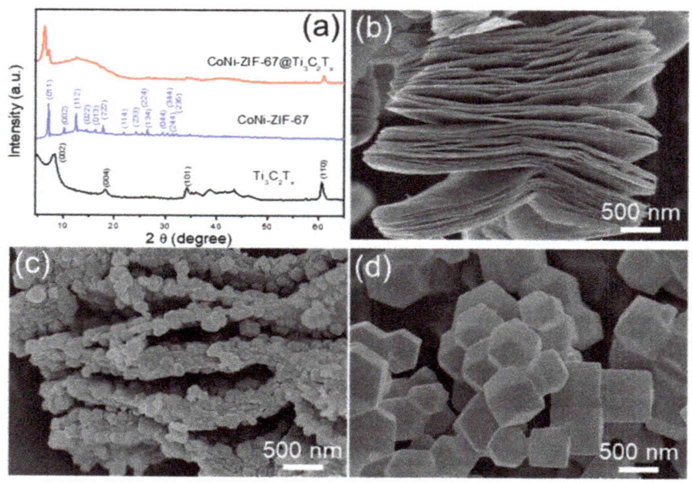

Figure 2. (**a**) X-ray diffraction (XRD) patterns of $Ti_3C_2T_x$, CoNi-ZIF-67@$Ti_3C_2T_x$, and pure CoNi-ZIF-67. Scanning electron microscopy (SEM) images of (**b**) $Ti_3C_2T_x$; (**c**) CoNi-ZIF-67@$Ti_3C_2T_x$, and (**d**) pure CoNi-ZIF-67.

The morphology of the as-prepared catalysts was characterized by SEM and TEM, revealing the hybrid structure of CoNi-ZIF-67@$Ti_3C_2T_x$. Figure 2b shows the SEM image of the pristine $Ti_3C_2T_x$, showing the typical accordion-like structure. The CoNi-ZIF-67@$Ti_3C_2T_x$ hybrid remains a multilayered structure as the pristine $Ti_3C_2T_x$ but attached with numerous small particles with a size of 100–200 nm between the interlayers of $Ti_3C_2T_x$ (Figure 2c). During the HF etching process, Al layers were removed from the MAX phase, and Ti atoms were bonded with the surface functional groups (–O, –OH, or –F), conferring the $Ti_3C_2T_x$ with negatively charged surfaces [23], which would facilitate the absorption of positively charged Co^{2+} and/or Ni^{2+} ions and subsequently coordinate with 2-methylimidazole molecules for an in-situ synthesis of the CoNi-ZIF-67 particles [11]. As a result, a significant intercalation of the $Ti_3C_2T_x$ layers is observed in Figure 2c due to the inclusion of the CoNi-ZIF-67 particles. Figure 2d presents the SEM image of the pure CoNi-ZIF-67, showing the typical rhombic dodecahedral structure of CoNi-ZIF-67 and a uniform particle size of 400–600 nm. Particularly, the CoNi-ZIF-67 particles grown on the $Ti_3C_2T_x$ surface are much smaller than the pure CoNi-ZIF-67 particles (Figure 2c,d). It is speculated that the $Ti_3C_2T_x$ matrix reduced the aggregation of CoNi-ZIF-67 and deterred the particle growth of CoNi-ZIF-67. In the preparation process, CTAB was added as a sealing agent to aid the formation of smaller and uniform CoNi-ZIF-67 particles. Figure S2 shows the SEM images of two CoNi-ZIF-67 catalysts with CTAB and without CTAB. It can be seen that the CoNi-ZIF-67 particles using CTAB are in the range of 400–600 nm (Figure S2a,b), while the CoNi-ZIF-67 without CTAB exhibits a relatively wider particle size range, from 300 nm to 1 μm (Figure S2c,d). Besides, it is revealed that the CTAB could also facilitate the intercalation of the CoNi-ZIF-67 particles into the interlayers of $Ti_3C_2T_x$ (Figure S3).

The TEM images in Figure 3 confirm the hybrid structure of CoNi-ZIF-67@$Ti_3C_2T_x$ and the rhombic dodecahedral structure of pure CoNi-ZIF-67. Compared with the accordion-like structure of $Ti_3C_2T_x$, the composite maintained the multilayered structure but was firmly attached with numerous particles on the surface and between the interlayers of $Ti_3C_2T_x$. It should be noted that the pristine MXene shows obvious lattice fringes for the layers (Figure 3b), while the lattice fringes were not observed in the CoNi-ZIF-67@$Ti_3C_2T_x$ (Figure 3d). We speculate that the MXene is coated with a thick layer of CoNi-ZIF-67 in the hybrid, and that it is therefore not as easy to observe the lattice fringes in the hybrid as in the pristine MXene. In addition, the high-angle annular dark-field scanning transmission electron microscopy (HAADF-STEM) and EDX elemental mapping images of the CoNi-ZIF-67@$Ti_3C_2T_x$ hybrid demonstrate the distribution of C, Co and Ni elements on the surface of MXene (Figure S4).

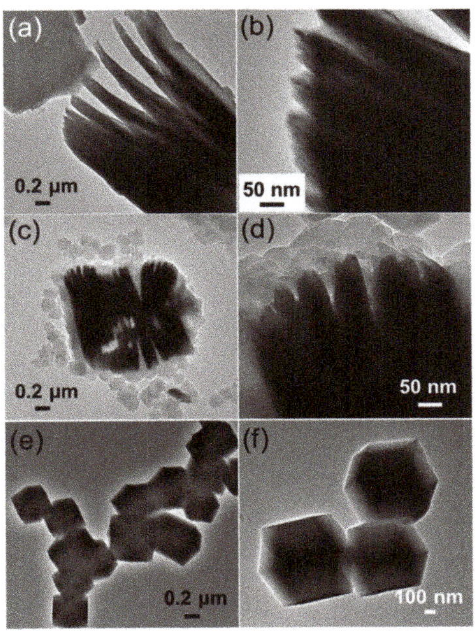

Figure 3. Transmission electron microscopy (TEM) images of (**a**,**b**) $Ti_3C_2T_x$, (**c**,**d**) CoNi-ZIF-67@$Ti_3C_2T_x$, and (**e**,**f**) pure CoNi-ZIF-67 at different magnifications.

The porosity of the as-prepared catalysts was measured by nitrogen adsorption isotherms, as shown in Figure S5a. The specific surface areas (S_{BET}) for $Ti_3C_2T_x$, CoNi-ZIF-67@$Ti_3C_2T_x$, and pure CoNi-ZIF-67 were 14.1, 202.9, and 1135.8 $m^2 \cdot g^{-1}$, respectively. The pure CoNi-ZIF-67 exhibits a dominant pore size of 1.08 and 1.3 nm, while the CoNi-ZIF-67@$Ti_3C_2T_x$ provides a larger dominant pore size of 1.74 nm (Figure S5b–d).

The XPS analysis further confirms the co-existence of C, Ti, Co and Ni elements in the CoNi-ZIF-67@$Ti_3C_2T_x$, with the elemental contents of 50.4, 11.2, 2.9 and 0.3 at.%, respectively (Table S1). The high resolution C 1s spectrum in Figure 4a can be deconvoluted into four peaks at 282.0, 285.0, 285.5 and 286.5 eV, which are attributed to the C–Ti, C=C, C–C and C–O species [11], respectively. The Ti region shows two pairs of $2p_{3/2}/2p_{1/2}$ doublets for the Ti–C (455.6 eV) and Ti–O (457.3 eV) species [21]. The Co 2p spectrum features three prominent species: Co^{2+} (782.6 eV for $2p_{3/2}$), Co^{3+}

(781.5 eV for $2p_{3/2}$) and satellite (787.2 eV) [24]. The MXene has an abundant number of surface termination groups (e.g., –O, –OH, and –F), which could adsorb the Co^{2+}/Ni^{2+} ions on the MXene surface and may change the Co/Ni oxidation during the pyrolysis process in the inert atmosphere. Consequently, the high-resolution Co/Ni XPS fitting may be helpful for explaining this part. The noise of the Co/Ni region is relatively high due to their low concentrations (Figure 4c,d). Consequently, we fitted the XPS data within the fitting error ($\sum \chi^2$) below 2. The Ni 2p region was analyzed into Ni^{2+} (855.0 eV for $2p_{3/2}$), Ni^{3+} (856.7 eV for $2p_{3/2}$) and satellite (861.1 eV) [25]. The core level peak analyses for the Co and Ni elements were listed in Tables S2 and S3. Interestingly, the CoNi-ZIF-67@$Ti_3C_2T_x$ exhibits a relatively higher ratio for the Co^{3+}/Co^{2+} species than the pure CoNi-ZIF-67 does, and the same trend can be observed in the Ni elements (Figure S6). We speculate that the introduction of MXene leads to the oxidation of the Co and Ni species in the CoNi-ZIF-67 phases, which may result from the numerous surface terminations on the MXene (–O or –OH), and which thus indicates the interaction between the MXene substrate and the in-situ grown CoNi-ZIF-67 phases.

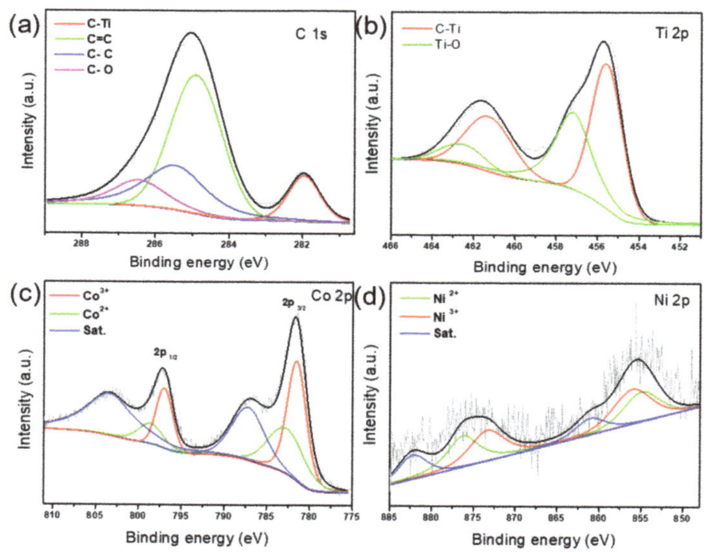

Figure 4. High resolution X-ray photoelectron spectroscopy (XPS) spectrum of (**a**) C 1s; (**b**) Ti 2p; (**c**) Co 2p and (**d**) Ni 2p for CoNi-ZIF-67@$Ti_3C_2T_x$.

The electrocatalytic activity of the as-prepared catalysts was first evaluated in a 0.1 M KOH solution in a standard three-electrode cell. Figure 5a presents the iR-corrected linear sweep voltammetry (LSV) curves at a scan rate of 50 $mV \cdot s^{-1}$. It is apparent that the $Ti_3C_2T_x$ has no OER activity. Meanwhile, the CoNi-ZIF-67@$Ti_3C_2T_x$ hybrid shows an enhanced electrocatalytic activity with a much larger current density than the pure CoNi-ZIF-67, which confirms the positive effect of the $Ti_3C_2T_x$ matrix on enhancing the OER activity. Accordingly, the CoNi-ZIF-67@$Ti_3C_2T_x$ displays a lower onset overpotential of 275 mV than the pure CoNi-ZIF-67 does (341 mV). The OER activity is also better than that of the as-purchased IrO_2 catalyst, with an onset potential of 281 mV, which indicates the good electrocatalytic performance of the CoNi-ZIF-67@$Ti_3C_2T_x$ hybrid. Another critical indicator of the OER activity is the overpotential at a current density of 10 $mA \cdot cm^{-2}$ ($\eta_{j=10}$), which is generally attributed to an approximately 10% efficient solar-to-fuel conversion device [23]. As listed in Figure 5b, the CoNi-ZIF-67@$Ti_3C_2T_x$ shows the lowest $\eta_{j=10}$ value (323 mV), when compared to the CoNi-ZIF-67 (389 mV) and IrO_2 catalysts (345 mV). It can be seen that the pure CoNi-ZIF-67 exhibited a poor catalytic performance, when compared to the as-purchased IrO_2 catalyst, which is mainly related to the instinct poor conductivity of the CoNi-ZIF-67, and which thus demonstrates the contribution of the MXene matrix to the good OER activity in the composite. Besides, in view of the XPS analysis (Tables S1–S3), no obvious changes in the Co/Ni atomic ratio, other than an apparent increase in the average oxidation state of both the Co and Ni elements, were detected after introducing the MXene matrix in the CoNi-ZIF-67 phase. We speculated that the enhanced OER activity of CoNi-ZIF-67@$Ti_3C_2T_x$ may also be related to the altering of the oxidation state of the transmission metal (Co and Ni) active sites [23].

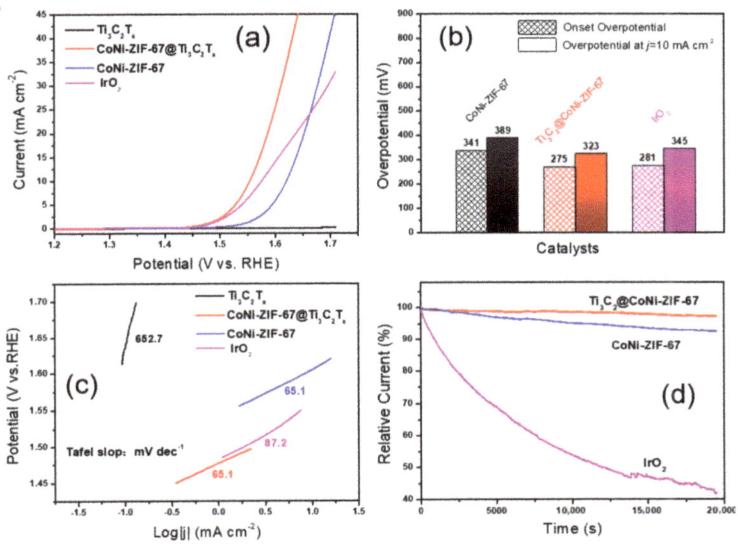

Figure 5. (**a**) Linear sweep voltammetry (LSV) curves of the catalyst $Ti_3C_2T_x$, CoNi-ZIF-67@$Ti_3C_2T_x$, pure CoNi-ZIF-67 and IrO_2 at 50 mV·s^{-1}; (**b**) A comparison of the catalysts in the onset potential and overpotential at a current density of 10 mA·cm^{-2}; (**c**) Tafel plots of the different catalysts; and (**d**) Chronoamperometry curves at 1.46 V vs. RHE over a 20,000 s continuous time.

The Tafel slope is a pivotal parameter for providing insightful information on the OER mechanism, particularly for the elucidation of OER kinetics and the rate-determining step [1]. In this regard, the Tafel slopes of the catalysts were plotted in Figure 5c. The value for CoNi-ZIF-67@$Ti_3C_2T_x$ is 65.1 mV·dec^{-1}, much smaller than that of the as-purchased IrO_2 catalyst (87.2 mV·dec^{-1}), thus revealing the higher OER rate and favorable kinetics of the CoNi-ZIF-67@$Ti_3C_2T_x$ hybrid. However, it should be note that, in our work, the Tafel plot of the as-purchased IrO_2 catalyst is higher than that of the reported nano-sized IrO_2 catalyst [26], which may be related to the morphology or size of the as-purchased IrO_2 in the reported works. Additionally, a comparison of the OER performance between the recently reported CoNi-based electrocatalysts with the CoNi-ZIF-67@$Ti_3C_2T_x$ in this work was listed in Table S4, indicating the excellent electrocatalytic properties of the CoNi-ZIF-67@$Ti_3C_2T_x$.

Furthermore, the durability of the catalysts was also performed at a constant potential of 1.46 V vs. RHE. As shown in the chronoamperometry curves (Figure 5d), the current of the CoNi-ZIF-67@$Ti_3C_2T_x$ hybrid remains nearly constant, with up to a 97.3% retention over a continuous time of 20,000 s, which is much more superior than that of the pure CoNi-ZIF-67 (92.6% retention) and that of the IrO_2 catalyst (only 43.4% remained), demonstrating the excellent stability of the CoNi-ZIF-67@$Ti_3C_2T_x$.

To better understand the catalytic activity of the CoNi-ZIF-67@$Ti_3C_2T_x$ hybrid, the electrochemically active surface area (ECSA) of the catalysts was investigated. The ECSA is normally positively correlated with the electrochemical double-layer capacitance (C_{dl}) [27]. Therefore, C_{dl} was calculated from the cyclic voltammetry (CV) curves at different scan rates in a narrow potential range of 1.023–1.073 V vs. RHE (Figure S7). As shown in Figure 6a, CoNi-ZIF-67@$Ti_3C_2T_x$ gives a much higher C_{dl} (5.77 mF·cm^{-2}) than the $Ti_3C_2T_x$ (1.18 mF·cm^{-2}) and the pure CoNi-ZIF-67 (1.57 mF·cm^{-2}) do, indicating a higher ECSA and more active sites in the hybrid. The higher ECSA is consistent with the smaller particle size of CoNi-ZIF-67 in the hybrid, as shown in Figure 2, which would expose more electrochemical active sites with the electrolyte solution. Additionally, the electrochemical impedance spectroscopy measurement was carried out to analyze the interfacial resistance of the electrocatalysts (Figure 6b). The Nyquist plots of the catalysts were fitted by the RC circuit model, as shown in the inset of Figure 6b, including an internal resistance (R1) and a charge transfer resistance (R2) for the electrochemical reaction [28–32]. The simulated R1 and R2 were shown in Table S5, which reveals that CoNi-ZIF-67@$Ti_3C_2T_x$ exhibits a smaller R1 and R2 than the pure CoNi-ZIF-67 and IrO_2, disclosing the optimized charge-transfer capacity of the hybrid during the OER process.

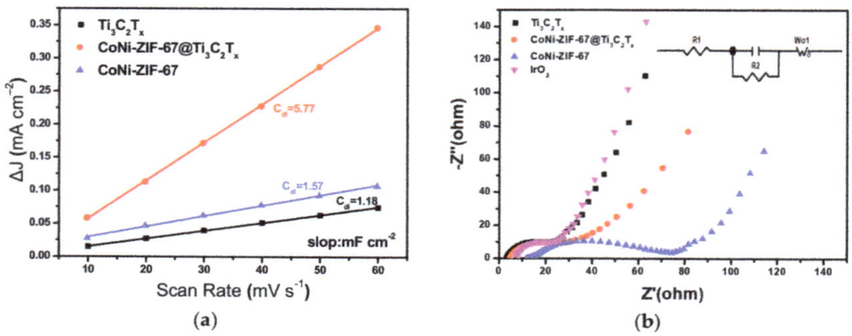

Figure 6. (**a**) $\Delta J = (Ja - Jc)$ plotted scan rates and (**b**) Nyquist plots for different catalysts.

4. Conclusions

In summary, a MXene supported CoNi-ZIF-67 hybrid was synthesized via the in-situ growth of CoNi-ZIF-67 rhombic dodecahedrons on the $Ti_3C_2T_x$ matrix via a coprecipitation reaction. It is shown that the addition of CTAB during the preparation process would aid the formation of smaller and uniform CoNi-ZIF-67 particles, while the CTAB could also facilitate the intercalation of CoNi-ZIF-67 particles into the interlayers of $Ti_3C_2T_x$, forming a hybrid structure composed of two phases. This CoNi-ZIF-67@$Ti_3C_2T_x$ hybrid exhibited a superior OER catalytic activity with a low onset potential (275 mV vs. a reversible hydrogen electrode, RHE) and Tafel slope (65.1 mV·dec^{-1}), much better than that of the IrO_2 catalysts and the pure CoNi-ZIF-67. On the basis of a comprehensive analysis, it is speculated that the good OER activity for the CoNi-ZIF-67@$Ti_3C_2T_x$ hybrid may be attributed to the following factors: (i) an enhanced conductivity of CoNi-ZIF-67 after the inclusion of the MXene matrix, (ii) a hybrid structure with smaller CoNi-ZIF-67 particles, (iii) an increase in the oxidation state of the Co and Ni elements after the introduction of MXene, and (iv) a high electrochemically active surface area for the hybrid. Therefore, the effective synergy shows the hybrid to be an excellent OER electrocatalyst that may pave a new way for the development of efficient non-precious metal electrocatalysts for OER.

Supplementary Materials: The following are available online at http://www.mdpi.com/2079-4991/9/5/775/s1, Table S1: Elemental compositions of Catalysts (at.%) determined by XPS, Table S2: Co2p core level peak analyses of catalysts (at.%), Table S3: Ni2p core level peak analyses of catalysts (at.%), Table S4: Comparisons of OER performance between recent reported CoNi-based electrocatalysts with CoNi-ZIF-67@$Ti_3C_2T_x$, Table S5, The simulated internal resistance (R1) and charge transfer resistance (R2) from the Nyquist plots, Figure S1: Enlarged image of XRD patterns of catalysts, Figure S2: SEM images of pure CoNi-ZIF-67 prepared by the same procedure but with CTAB (a,b) and without CTAB (c,d), Figure S3: SEM image of CoNi-ZIF-67@$Ti_3C_2T_x$ without using CTAB, Figure S4: HAADF-STEM images and the corresponding elemental maps of C, Ti, Co and Ni in the CoNi-ZIF-67@$Ti_3C_2T^x$, Figure S5: Nitrogen adsorption isotherms and pore size distribution of catalysts, Figure S6:

XPS results of $Ti_3C_2T_x$ and CoNi-ZIF-67, Figure S7: CV curves in a potential range of 1.023–1.073 V vs. RHE of catalysts.

Author Contributions: Y.W. designed the experiments and wrote the paper; Z.L. designed the experiments; Z.W., C.M. and X.X. performed the experiments and analyzed the data, D.L. conducted the work and revised the paper.

Funding: This research was financially supported by Science Foundation of China University of Petroleum, Beijing No. 2462017YJRC013 (Y.W.), No. 2462019BJRC001 (Z.L.), No. 2462018BJB002 (D.L.), Beijing Natural Science Foundation (Grant No. 2182061) (Z.L.) and Beijing Municipal Natural Science Foundation No. 2184119 (D.L.).

Conflicts of Interest: The authors declare no conflict of interest.

References

1. Suen, N.T.; Hung, S.F.; Quan, Q.; Zhang, N.; Xu, Y.J.; Chen, H.M. Electrocatalysis for the oxygen evolution reaction: Recent development and future perspectives. *Chem. Soc. Rev.* **2017**, *46*, 337–365. [CrossRef] [PubMed]
2. Osgood, H.; Devaguptapu, S.V.; Xu, H.; Cho, J.; Wu, G. Transition metal (Fe, Co, Ni, and Mn) oxides for oxygen reduction and evolution bifunctional catalysts in alkaline media. *Nano Today* **2016**, *11*, 601–625. [CrossRef]
3. Ma, T.Y.; Dai, S.; Jaroniec, M.; Qiao, S.Z. Metal-organic framework derived hybrid Co_3O-carbon porous nanowire arrays as reversible oxygen evolution electrodes. *J. Am. Chem. Soc.* **2014**, *136*, 13925–13931. [CrossRef]
4. Zhao, S.; Wang, Y.; Dong, J.; He, C.T.; Yin, H.; An, P.; Zhao, K.; Zhang, X.; Gao, C.; Zhang, L.; et al. Ultrathin metal-organic framework nanosheets for electrocatalytic oxygen evolution. *Nat. Energy* **2016**, *1*, 16184. [CrossRef]
5. Zhao, Q.; Yan, Z.; Chen, C.; Chen, J. Spinels: Controlled preparation, oxygen reduction/evolution reaction application, and beyond. *Chem. Rev.* **2017**, *117*, 10121–10211. [CrossRef] [PubMed]
6. Qian, H.; Tang, J.; Wang, Z.; Kim, J.; Kim, J.H.; Alshehri, S.M.; Yanmaz, E.; Wang, X.; Yamauchi, Y. Synthesis of cobalt sulfide/sulfur doped carbon nanocomposites with efficient catalytic activity in the oxygen evolution reaction. *Chem. Eur. J.* **2016**, *22*, 18259–18264. [CrossRef] [PubMed]
7. Yu, X.Y.; Feng, Y.; Guan, B.; Lou, X.W.; Paik, U. Carbon coated porous nickel phosphides nanoplates for highly efficient oxygen evolution reaction. *Energy Environ. Sci.* **2016**, *9*, 1246–1250. [CrossRef]
8. Morozan, A.; Jaouen, F. Metal organic frameworks for electrochemical applications. *Energy Environ. Sci.* **2012**, *5*, 9269–9290. [CrossRef]
9. Lee, J.; Farha, O.K.; Roberts, J.; Scheidt, K.A.; Nguyen, S.T.; Hupp, J.T. Metal-organic framework materials as catalysts. *Chem. Soc. Rev.* **2009**, *38*, 1450–1459. [CrossRef] [PubMed]
10. Liu, S.; Wang, Z.; Zhou, S.; Yu, F.; Yu, M.; Chiang, C.Y.; Zhou, W.; Zhao, J.; Qiu, J. Metal-organic-framework-derived hybrid carbon nanocages as a bifunctional electrocatalyst for oxygen reduction and evolution. *Adv. Mater.* **2017**, *29*, 1700874. [CrossRef]
11. Zhao, L.; Dong, B.; Li, S.; Zhou, L.; Lai, L.; Wang, Z.; Zhao, S.; Han, M.; Gao, K.; Lu, M.; et al. Interdiffusion reaction-assisted hybridization of two-dimensional metal-organic frameworks and $Ti_3C_2T_x$ nanosheets for electrocatalytic oxygen evolution. *ACS Nano* **2017**, *11*, 5800–5807. [CrossRef]
12. Wang, L.; Wu, Y.; Cao, R.; Ren, L.; Chen, M.; Feng, X.; Zhou, J.; Wang, B. Fe/Ni metal-organic frameworks and their binder-free thin films for efficient oxygen evolution with low overpotential. *ACS Appl. Mater. Interfaces* **2016**, *8*, 16736–16743. [CrossRef]
13. Naguib, M.; Kurtoglu, M.; Presser, V.; Lu, J.; Niu, J.; Heon, M.; Hultman, L.; Gogotsi, Y.; Barsoum, M.W. Two-dimensional nanocrystals produced by exfoliation of Ti_3AlC_2. *Adv. Mater.* **2011**, *23*, 4248–4253. [CrossRef] [PubMed]
14. Naguib, M.; Mochalin, V.N.; Barsoum, M.W.; Gogotsi, Y. 25th anniversary article: MXenes: A new family of two-dimensional materials. *Adv. Mater.* **2014**, *26*, 992–1005. [CrossRef]
15. Mashtalir, O.; Lukatskaya, M.R.; Zhao, M.Q.; Barsoum, M.W.; Gogotsi, Y. Amine-assisted delamination of Nb_2C MXene for li-ion energy storage devices. *Adv. Mater.* **2015**, *27*, 3501–3506. [CrossRef] [PubMed]
16. Xiong, D.; Li, X.; Bai, Z.; Lu, S. Recent advances in layered $Ti_3C_2T_x$ MXene for electrochemical energy storage. *Small* **2018**, *14*, 1703419. [CrossRef] [PubMed]

17. Ran, J.; Gao, G.; Li, F.T.; Ma, T.Y.; Du, A.; Qiao, S.Z. Ti_3C_2 MXene co-catalyst on metal sulfide photo-absorbers for enhanced visible-light photocatalytic hydrogen production. *Nat. Commun.* **2017**, *8*, 13907. [CrossRef]
18. Zhang, Q.; Teng, J.; Zou, G.; Peng, Q.; Du, Q.; Jiao, T.; Xiang, J. Efficient phosphate sequestration for water purification by unique sandwich-like MXene/magnetic iron oxide nanocomposites. *Nanoscale* **2016**, *8*, 7085–7093. [CrossRef]
19. Shahzad, F.; Alhabeb, M.; Hatter, C.B.; Anasori, B.; Hong, S.M.; Koo, C.M.; Gogotsi, Y. Electromagnetic interference shielding with 2D transition metal carbides (MXenes). *Science* **2016**, *353*, 1137–1140. [CrossRef]
20. Li, Z.; Zhuang, Z.; Lv, F.; Zhu, H.; Zhou, L.; Luo, M.; Zhu, J.; Lang, Z.; Feng, S.; Chen, W.; et al. The marriage of the FeN_4 moiety and MXene boosts oxygen reduction catalysis: Fe 3d electron delocalization matters. *Adv. Mater.* **2018**, *30*, 1803220. [CrossRef]
21. Wen, Y.; Rufford, T.E.; Chen, X.; Li, N.; Lyu, M.; Dai, L.; Wang, L. Nitrogen-doped $Ti_3C_2T_x$ MXene electrodes for high-performance supercapacitors. *Nano Energy* **2017**, *38*, 368–376. [CrossRef]
22. Ammar, M.; Jiang, S.; Ji, S. Heteropoly acid encapsulated into zeolite imidazolate framework (ZIF-67) cage as an efficient heterogeneous catalyst for Friedel-Crafts acylation. *J. Solid State Chem.* **2016**, *233*, 303–310. [CrossRef]
23. Yu, M.; Zhou, S.; Wang, Z.; Zhao, J.; Qiu, J. Boosting electrocatalytic oxygen evolution by synergistically coupling layered double hydroxide with MXene. *Nano Energy* **2018**, *44*, 181–190. [CrossRef]
24. Wang, C.; Zhu, X.D.; Mao, Y.C.; Wang, F.; Gao, X.T.; Qiu, S.Y.; Le, S.R.; Sun, K.N. MXene-supported Co_3O_4 quantum dots for superior lithium storage and oxygen evolution activities. *Chem. Commun.* **2019**, *55*, 1237–1240. [CrossRef]
25. Tang, Y.; Yang, C.; Yang, Y.; Yin, X.; Que, W.; Zhu, J. Three dimensional hierarchical network structure of S-$NiFe_2O_4$ modified few-layer titanium carbides (MXene) flakes on nickel foam as a high efficient electrocatalyst for oxygen evolution. *Electrochim. Acta* **2019**, *296*, 762–770. [CrossRef]
26. Abbott, D.; Lebedev, D.; Waltar, K.; Povia, M.; Nachtegaal, M.; Fabbri, E.; Coperet, C.; Schmidt, T. Iridium oxide for the oxygen evolution reaction: Correlation between particle size, morphology, and the surface hydroxo layer from Operando XAS. *Chem. Mater.* **2016**, *28*, 6591–6604. [CrossRef]
27. He, P.; Yu, X.Y.; Lou, X.W. Carbon-incorporated nickel-cobalt mixed metal phosphide nanoboxes with enhanced electrocatalytic activity for oxygen evolution. *Angew. Chem. Int. Ed.* **2017**, *56*, 3897–3900. [CrossRef]
28. Chauhan, M.; Reddy, K.P.; Gopinath, C.S.; Deka, S. Copper cobalt sulfide nanosheets realizing a promising electrocatalytic oxygen evolution reaction. *ACS Catal.* **2017**, *7*, 5871–5879. [CrossRef]
29. Wen, Y.; Rufford, T.E.; Hulicova-Jurcakova, D.; Wang, L. Nitrogen and phosphorous co-doped graphene monolith for supercapacitors. *ChemSusChem* **2016**, *9*, 513–520. [CrossRef]
30. Li, H.; Yang, J.; Gao, M.; Wang, J.; Sun, B. Washing rice before cooking has no large effect on the texture of cooked rice. *Food Chem.* **2019**, *271*, 388–392. [CrossRef]
31. Wen, Y.; Rufford, T.E.; Hulicova-Jurcakova, D.; Zhu, X.; Wang, L. Structure control of nitrogen-rich graphene nanosheets using hydrothermal treatment and formaldehyde polymerization for supercapacitors. *ACS Appl. Mater. Interfaces* **2016**, *8*, 18051–18059. [CrossRef] [PubMed]
32. Wen, Y.; Wang, B.; Huang, C.; Wang, L.; Hulicova-Jurcakova, D. Synthesis of phosphorus-doped graphene and its wide potential window in aqueous supercapacitors. *Chem. Eur. J.* **2015**, *21*, 80–85. [CrossRef]

nanomaterials

MDPI

Article

Facile Synthesis of Magnetic Nitrogen-Doped Porous Carbon from Bimetallic Metal–Organic Frameworks for Efficient Norfloxacin Removal

Hui Wang [1], Xi Zhang [2], Yan Wang [1], Guixiang Quan [1], Xiangyun Han [1] and Jinlong Yan [1,*]

1 School of Environmental Science and Engineering, Yancheng Institute of Technology, Yancheng 224051, China; whsl@ycit.cn (H.W.); shanliangbang@163.com (Y.W.); qgx@ycit.cn (G.Q.); hxy16_2000@163.com (X.H.)

2 College of Life and Environmental Science, Shanghai Normal University, Shanghai 200234, China; m18321255826_1@163.com

* Correspondence: yjlyt@ycit.cn; Tel.: +86-515-8829-8805

Received: 6 August 2018; Accepted: 23 August 2018; Published: 26 August 2018

Abstract: Magnetic nitrogen-doped porous carbon (MNPC) has been prepared via self-catalytic pyrolysis of bimetallic metal-organic frameworks (MOFs). The as-obtained MNPC showed favorable features for antibiotics adsorption such as high specific surface area (871 m^2 g^{-1}), high pore volume (0.75 cm^3 g^{-1}), porous structure, good graphitization degree, and rich N-doping. Moreover, the MNPC has magnetic properties due to the Co species, which is embedded with a high dispersion, so the absorbent can be easily separated. Based on the above excellent characteristics, the MNPC was used as the absorbent for norfloxacin (NOR) removal. The experimental maximum NOR adsorption capacity of MNPC was 55.12 mg g^{-1} at 298.15 K and a pH of 6.0 with an initial NOR concentration of 50 mg L^{-1}. The data analysis of the kinetics revealed that the experimental data of NOR uptakes versus time agreed with the pseudo-second order model. The isotherm data analysis revealed the favorable application of the Freundlich model. Based on the adsorption results over a wide range of conditions, the dominant adsorption mechanisms were found to be pore-filling, electrostatic interaction, and the H-bond.

Keywords: self-catalytic pyrolysis; porous carbon; metal–organic frameworks; antibiotics; adsorption

1. Introduction

Over the past few decades, the emission of pharmaceutical compounds into the environment has sharply increased due to fast population growth and the rapid expansion of the pharmaceuticals industry. Antibiotics are one of most important type of pharmaceuticals, and are usually used as drugs or feed additives [1–4]. However, large amounts of antibiotics are stable and cannot be easily degraded, thus they are persistent in the environment. In addition, antibiotics could generate antibiotic-resistance genes in microorganisms, which can proliferate and widely disseminate in ecosystems. Fluoroquinolones are a commonly used antibiotic and their concentration is relatively high in the environment [5,6]. Norfloxacin (NOR) is one of the most frequently used fluoroquinolone antibiotics, and is always used to treat infectious diseases. It has been detected in the surface water and found to be toxic to aquatic organisms and human beings [7,8]. Therefore, it is necessary to develop a cost-effective method to remove NOR from wastewater.

Such methods as advanced oxidation, electrochemical methods, and biological treatments have been applied extensively to remove NOR from wastewaters. Among all of these methods, adsorption is the top priority owing to its simplicity, low operating cost, safety, and efficiency [9–11]. With such advantages as a large specific surface area and porous structure, carbon materials have been applied to remove NOR from water. For example, Xing et al. investigated the adsorption of norfloxacin (NOR) onto multiwall carbon nanotubes and activated carbon, and the results showed that activated carbon

(AC) has a better NOR sorption capacity due to its higher surface area [12]. Theydan et al. prepared AC from a lignocellulosic biomass to remove NOR from water, and a maximum removal percentage of 98.13% was achieved [13]. Although a significant amount of research has been expended on adsorbing materials for antibiotics removal during the past few decades, development of novel adsorbents with higher performance is still needed.

As a new class of porous inorganic–organic materials, metal–organic frameworks (MOFs), have attracted wide attention owing to their high surface area and tunable pore size, which is widely used in areas of drug delivery, gas storage, and separation and catalysis [14–17]. Recently, MOFs have been used as templates or novel sources to prepare porous carbons through further carbonization. For example, Xu et al. applied MOFs as sacrificial templates to synthesis nanoporous carbons for the first time. They introduced furfuryl alcohol into the MOF-5 through a vapor phase protocol, which was then carbonized at 1000 °C under an inert atmosphere to obtain porous carbon [18]. Park et al. presented hierarchically porous carbon from highly crystalline MOFs and used it as a hydrogen storage adsorbent [19]. Most recently, Huang et al. demonstrated the application of MOF-derived porous carbon as an adsorbent for antibiotics removal. They prepared porous carbon through a one-step carbonization of zeolitic imidazolate framework-8 (ZIF-8), which showed a larger specific surface area due to the vaporization of the center metal (Zn) of ZIF-8 during the pyrolysis process. They were further used for ciprofloxacin removal from water [20]. Although porous carbon derived from ZIF-8 has a high specific surface area for antibiotics adsorption, it has limitations in terms of adsorption capacity and ease of separation.

Recently, the sustainability of the adsorption process has been advocated, such as: green adsorbent and green separation methods [21–23]. It has been demonstrated that the incorporation of magnetic nanoparticles on the surface of adsorbents can be engineered to allow the magnetic separation and recovery of the absorbents [24,25]. Wang et al. synthesized reduced graphene oxide/magnetite composites through an in situ reaction and utilized it as an adsorbent with a magnetically separable property for fluoroquinolone antibiotics [26]. Cai et al. encapsulated magnetic nanoparticles into carbon with a well-constructed core-shell structure, and then used it as an adsorbent for organic pollutants isolation [27]. However, the preparation of magnetic adsorbents usually needs additional processes to load the magnetic metal oxide, and it is difficult to control the dispersion of loaded particles during the synthesis process.

Herein, we report a simple but efficient solution process for the fabrication of a new form of magnetic nitrogen-doped porous carbon (MNPC) adsorbents for the NOR removal. The MNPC was directly synthesized by self-catalytic pyrolysis of bimetallic MOFs, which were prepared by using divalent Zn^{2+} and Co^{2+} as center metal ions and 2-methylimidazole as the ligand (Figure 1). In the carbonization process, the Zn, with a boiling point of around 900 °C, was evaporated during the calcination process, and the porous structure was formed simultaneously. Furthermore, the Co species were embedded in the porous structure with a high dispersion due to the coordination structure of the MOF's precursor, and so the MNPC had magnetic properties. Moreover, the Co species can act as catalyst to improve the graphitization degree of MNPC, which can enhance the adhesion between antibiotics and adsorbents through π–π conjugation. More importantly, by the development of such multiple structures, the adsorption performance was significantly enhanced.

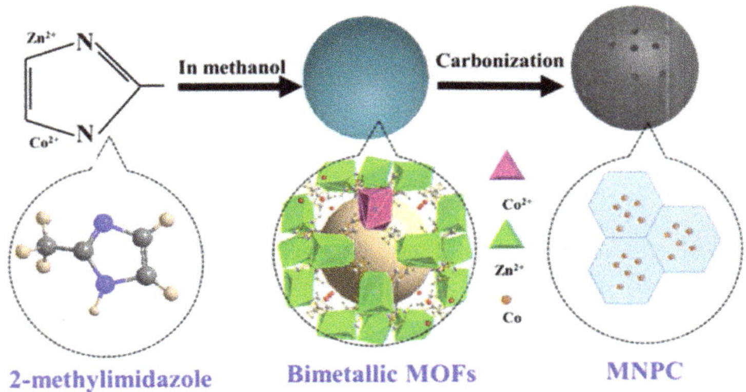

Figure 1. Schematic illustration of the construction process for the magnetic nitrogen-doped porous carbon (MNPC).

2. Materials and Methods

2.1. Synthesis of MNPC

Materials: The zinc nitrate hexahydrate ($Zn(NO_3)_2{\cdot}6H_2O$), cobaltous nitrate hexahydrate ($Co(NO_3)_2{\cdot}6H_2O$), 2-methylimidazole and norfloxacin were analytic grade provided by Aladdin Chemical Reagent Co., Ltd., Shanghai, China.

Synthesis of MNPC: Typically, $Co(NO_3)_2{\cdot}6H_2O$ (0.27 g, 0.9 mmol) and ($Zn(NO_3)_2{\cdot}6H_2O$) (1.40 g, 4.7 mmol) were first dissolved in 100 mL of methanol. 2-Methylimidazole (3.70 g, 45.1 mmol) in 100 mL methanol was then added to the above solution. After quickly stirring for 24 h, the products were separated by centrifugation and washed thoroughly with methanol. The obtained bimetallic MOFs were dried at 50 °C overnight, and further activated at 200 °C for 24 h under a vacuum before use. The as-synthesized bimetallic MOFs nanocrystals were heated to 950 °C with the ramp rate of 3 °C/min under a N_2 atmosphere and carbonized at 950 °C for 2 h, and then cooled to room temperature naturally. Finally, the MNPC was entirely fabricated. The magnetic carbon (MC) prepared by MOFs with only Co ions as a central ion was used for comparison. The porous carbon (PC) prepared by MOFs with only Zn ions as a central ion was also prepared. In fact, the methanol used in this process could be recycled through membranes to realize sustainable fabrication [28,29].

2.2. Adsorption Performance of MNPC

The adsorption experiments on NOR were conducted in 250 mL stopper conical flasks, and then placed in a thermostatic shaker with a speed of 200 rpm. In the adsorption experiment, 80 mg of adsorbent was added to 100 mL of adsorbate solution. The influence of initial concentrations (5–50 mg L^{-1}), pH (2–10), temperature and ionic strength on the adsorption of NOR were also investigated. The solution pH was adjusted by dilute HCl or NaOH solution. The concentration of NOR was measured with a UV–Visible spectrophotometer (TU-1810, Beijing Purkinje General Instrument Co. Ltd., Beijing, China) at 272 nm. The adsorbed capacity (q) and removal rate (η) were calculated according to the following equations:

$$q = (C_0 - C_t)V/m \quad (1)$$

$$\eta = (C_0 - C_t)/C_0 \quad (2)$$

where C_0 and C_t (mg L^{-1}) represent the initial and final concentrations of NOR in the feed solution, respectively, V is the volume of NOR solution (L), and m is the dry mass of MNPC (g).

3. Results

3.1. Characterization

The X-ray diffraction (XRD, Rigaku D/Max-RB, Rigaku Corporation, Tokyo, Japan) measurements is usually conducted to evaluate the structure of materials, and the XRD pattern of the MNPC is presented in Figure 2a. The MNPC shows an obvious diffraction peak at the 2θ = 26°, corresponding to the (002) plane of the graphitic carbon [30]. The diffraction peaks located at around 44° and 51° are ascribed to *fcc* Co, which is embedded in the carbon shell [31,32]. There was no characteristic peak of Zn in the XRD patterns due to the effective evaporation during the high-temperature calcination. The XRD pattern of MC was similar to that of MNPC (Figure S1). The graphitization degree of MNPC was further detected using Raman spectra (JY H800UV, Jobin-Yvon Corporation, Longjumeau, France), and the result is shown in Figure 2b. Two broad peaks at 1330 and 1583 cm^{-1} are obvious, and are related to the D-band and G-band, respectively. The D-band is associated with defects in the carbon structure, while the G-band is attributed to the vibration of sp^2 carbon atoms in both the rings and chain [25]. The graphitization degree of MNPC can be found by calculating the ratios of the integrated intensities of the graphitic G-band to that of the disorder-induced D-band. As calculated, the value of I_G/I_D was 1.07, which was close to the MC (1.02, Figure S2a). However, the values of I_G/I_D were higher than that of PC (0.94, Figure S2b) due to the catalytic action of Co.

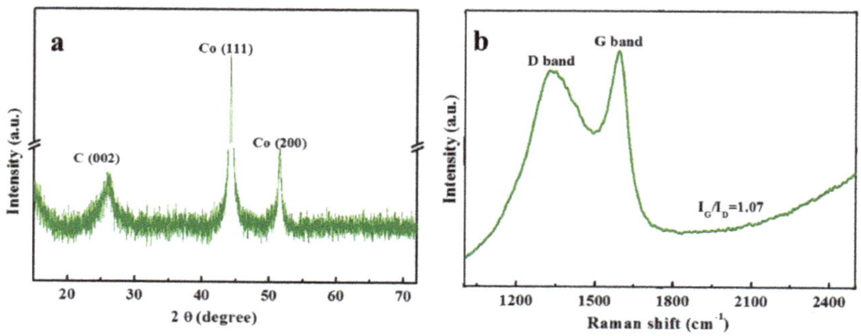

Figure 2. (**a**) X-ray diffraction (XRD) pattern and (**b**) Raman spectrum of the MNPC.

To further probe for the chemical identification of elements in the MNPC, the X-ray photoelectron spectroscopy (XPS, PHI-5000C ESCA system, Perkin–Elmer, Hopkinton, MA, USA) measurements were performed. According to the results, the elemental content of C, N, O, and Co was 80.29, 10.58, 7.19, 1.95 at %, respectively. The C, N, O, and Co contents of MC are 90.67, 3.6, 4.38, and 1.34 at %, respectively (Figure S3). The high-resolution C1s spectra (Figure 3a) could be fitted with three peaks at 284.6, 286.4, and 287.8 eV, corresponding to the sp^2 aromatic rings, C–O, and C=O, respectively [33]. The peak of sp^2 carbon showed the strongest intensity, indicating that the MNPC predominantly consisted of sp^2-hybridized carbon due to the effective catalytic graphitization. As seen from the high-resolution N1s (Figure 3b), three different types of nitrogen species were well deconvoluted. The N-6 atoms were located at 398.5 eV, and were bonded with two carbon atoms in a C_6 ring, so a pair of lone electrons could be introduced simultaneously. This was beneficial to the formation of a hydrogen bond with the NOR molecule. The N-5 was centered at 400.4 eV, associated with the adjacent phenolic or carbonyl group. The N–Q atoms bond with three carbon atoms in the center of graphitic plane [34,35]. The additional N-doping can increase the adsorption sites for NOR, and further improve the adsorption performance of MNPC.

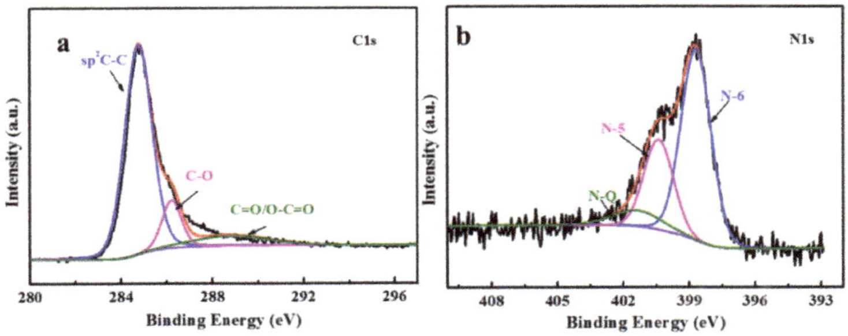

Figure 3. (**a**) C 1s spectra and (**b**) N 1s spectra of the MNPC.

The N_2 sorption isothermal (ASAP 2020, Micromeritics Inc., Norcross, GA, USA) was further examined to analyze the pore structure of MNPC. As seen from Figure 4a, the MNPC showed a typical IV-type isotherm with a hysteresis loop at p/p_0 = 0.4–1.0 (inset), indicating the mesoporous structure of MNPC [36]. Figure 4b shows the Barrett–Joyner–Halenda (BJH) pore size distribution profile of MNPC derived from desorption branches of isotherms. Clearly, the MNPC pores' radii was mainly concentrated at 2.0 nm, further indicating that the mesopores were dominant in the structure of MNPC. The specific surface area of MNPC was 871 $m^2\ g^{-1}$, much larger than that of MC obtained with the absence of a Zn ion (Figure S4). With Zn coordination, the ZnO would be formed during the carbonization process, which can act as sacrificial template accelerating the formation of the porous structure of MNPC [37,38]. Furthermore, the pore volume of MNPC was 0.76 $cm^3\ g^{-1}$ (Figure 4b), which is much larger than that of MC (0.16 $cm^3\ g^{-1}$). The increased specific surface area and pore volume are favorable for increasing the accessible surface area for NOR accumulation during the adsorption process and then enhance the adsorption capacity.

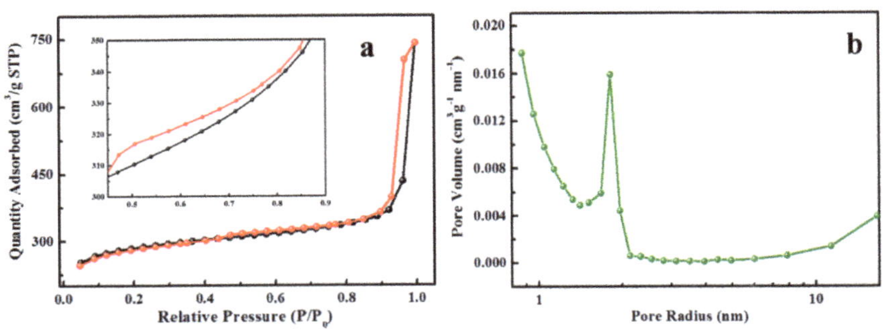

Figure 4. (**a**) N_2 sorption isotherm and (**b**) pore size distribution of the MNPC.

The surface morphology of MNPC was investigated with scanning electron microscopy (SEM, JEOL JSM-6700F, Tokyo, Japan) and transmission electron microscopy (TEM, JEOL JEM-200CX, Tokyo, Japan). As seen from the SEM image in Figure 5a, the bimetallic MOFs precursor shows a cubic-like structure with an average size of 50 nm. After pyrolysis at 950 °C, the Zn species were volatilized, and the pores left simultaneously [38]. Moreover, the MNPC retained the morphology of the MOF's precursor with a good dispersion (Figure 5b). This indicates that the structure kept well, even after the high-temperature calcination. The TEM image in Figure 5c reveals that the MNPC had a uniform morphology with an interconnected porous structure, and the Co nanopaticals (NPs) were embedded in the porous carbon. The high-resolution transmission electron microscopy (HRTEM) image (Figure 5d) shows further that MNPC exhibits an obvious core-shell structure. The graphitic carbon structures were the shells with an interplane spacing of (002) crystal lattice (3.4 Å), which resulted from the catalytic graphitization behavior of Co NPs [31]. Moreover, the Co NPs were tightly wrapped by graphitic carbon shells due to the coordinating structures of bimetallic MOFs as a precursor. In addition, the HRTEM image shows a distinct lattice fringe with an interplanar spacing of 0.2 nm, which matched well with the spacing of (111) planes of the Co phase. Besides, the MC showed a dodecahedron-like structure with a particle size around 250 nm, and the Co also embedded in the carbon structure (Figure S5).

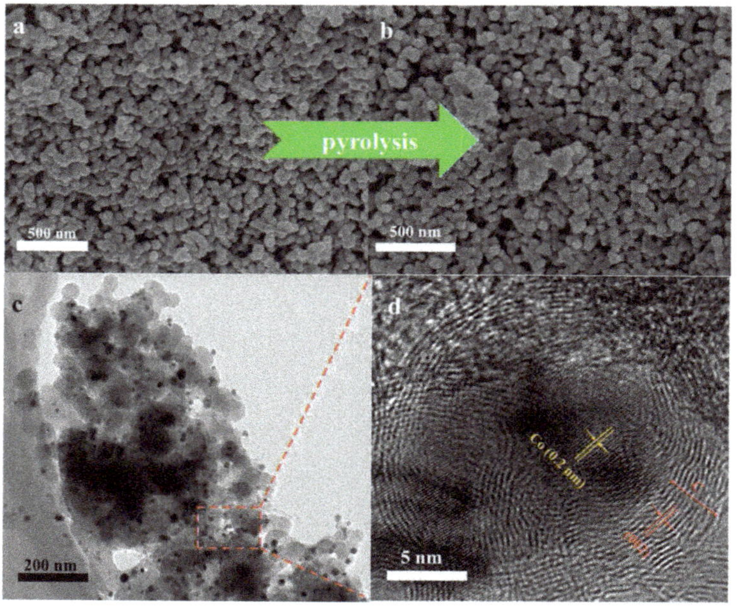

Figure 5. SEM images of (**a**) Bimetallic metal-organic frameworks (MOFs) precursor; and (**b**) the MNPC; (**c**) TEM and (**d**) HRTEM images of the MNPC.

As seen from the high-angle annular dark field-scanning electron microscopy (HAADF-STEM) (Figure 6a), The Co NPs were embedded into graphitic carbon structure. The elemental mapping was performed to illustrate the spatial distribution of C, N, O, and Co in the structure of MNPC in Figure 6b. As revealed in Figure 6c–f, the elemental mapping results further confirmed the uniform distribution of Co and N species within the MNPC structure. Besides, the MC also showed a homogeneous distribution (Figure S5). It is generally accepted that the N species can promote the formation of hydrogen bonds and accelerate the adsorption performance. Moreover, the Co species with a good dispersion within the carbon structure is beneficial to the further separation of adsorbents.

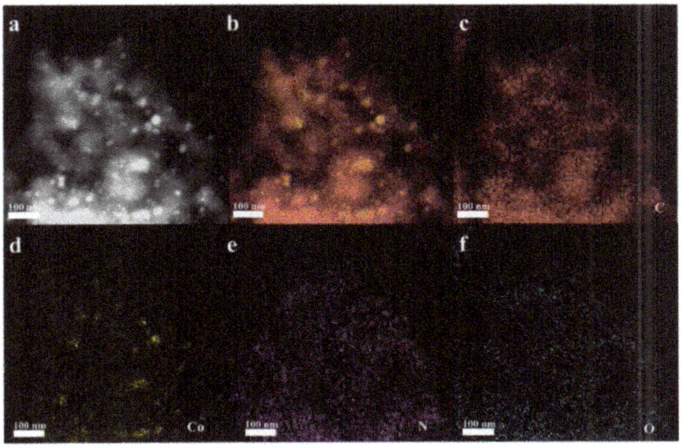

Figure 6. The high-angle annular dark field-scanning electron microscopy (HAADF-STEM): (**a**) image, and (**b–f**) mapping images of MNPC.

3.2. *Adsorption Performance*

The adsorption behavior of the MNPC on the NOR was investigated by batch mode experiments in 10 mg L^{-1} aqueous solution. As seen from the adsorption curves in Figure 7, the adsorption capacity sharply increased with the adsorption time, suggesting that the NOR in the aqueous solution could be quickly and easily removed by the adsorbents [39]. As time goes on, the change of adsorption capacity became slower until reaching an adsorption equilibrium owing to the fact that the number of adsorption sites decreased as the adsorption time increased. Obviously, the adsorption capacity of the MNPC adsorbents was much larger than that of MC, indicating that the MNPC exhibited much better adsorption performance. After 150 min, the final adsorption capacity of MNPC adsorbents was 8.84 mg g^{-1}, larger than that of MC (7.98 mg g^{-1}). As seen from the Figure 7b, the MNPC in the aqueous solution could be easily separated under an external magnetic field.

The better adsorption performance of MNPC adsorbents can be ascribed to the following aspects: First, the larger specific surface area of MNPC can provide more adsorption sites for the NOR adsorption. Second, the NOR molecules can be easily transported between the smooth channels in the MNPC due to the interconnected porous structure. Moreover, the MNPC has a good graphitization degree, which is beneficial to the formation of π–π interactions between NOR and absorbents, which then further improves the adsorption capacity [8,40]. Furthermore, hydrogen bonding is easy to form between MNPC adsorbents and NOR due to the effect of nitrogen doping, which further promotes adsorption performance [38]. In conclusion, combining the pore structure, large specific surface area, good graphitization degree, and the effective nitrogen doping, MNPC can be considered as an excellent candidate material for NOR adsorption application.

As is well known, the structure and surface properties of adsorbents have important influence on the adsorption performance. Generally, the adsorbent's structure has a great effect on the physical adsorption, and the chemical adsorption is usually related to the functional groups on the surface of adsorbents [41]. The MNPC has large specific surface area, which can provide abundant adsorption sites for NOR adsorption. The porous structure is beneficial to the NOR molecules' penetration. In addition, such oxygen-containing functional groups as –COOH and –OH and N-doping are on the surface of MNPC, so the hydrogen bonding can be easily formed between the NOR molecules the MNPC, which then promotes the adsorption capacity [8]. Moreover, the aromatic structures and C=C double bonds in NOR can contribute to the affinity between MNPC and NOR through the π–π interactions and then increase the adsorption capacity [42].

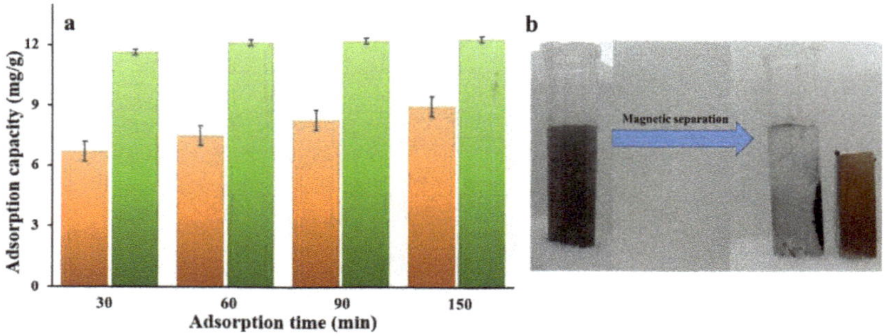

Figure 7. (**a**) Plots of adsorption capacity vs adsorption time of MNPC and MC in the NOR aqueous solutions at concentrations of 10 mg L^{-1}; and (**b**) the photo of MNPC separated under an external magnetic field.

As is well known, the amount of adsorbents has a critical effect on the adsorption performance. The influence of the adsorbent's dosage was explored by adding various amounts of MNPC to 100 mL of a 10 mg L^{-1} NOR solution. As seen from Figure 8a, the adsorption capacity decreased with the increase of the absorbent's dosage due to the completely exposed adsorption sites at the low dosage. While at higher dosage, the unoccupied adsorption sites were excess and resulted in a lower adsorption capacity [43]. Considering the adsorbent amounts and adsorption capacity, an 0.8 g/L dosage of MNPC was selected for further studies.

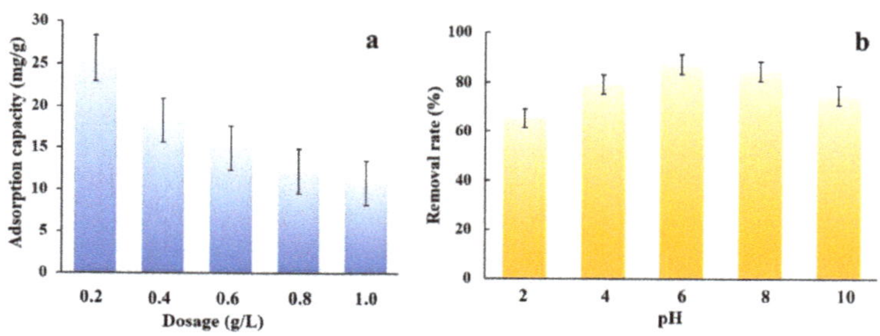

Figure 8. (**a**) Plots of adsorption capacity vs dosage; and (**b**) plots of removal rates vs. pH value with a dosage of 0.8 g L^{-1}. All the curves were obtained in a 10 mg L^{-1} NOR aqueous solution at 30 °C.

pH is another important factor affecting adsorption performance. As seen in Figure 8b, the NOR adsorption on MNPC initially increased with the pH value ranging from 2.0 to 6.0, and then decreased when the solution pH value was higher than 6.0. The NOR contained a carboxyl and piperazinyl group, which shows two proton-binding sites. Its two acid dissociation constant pKa values were 6.22 and 8.51, respectively. In the solution, the protonation–deprotonation reaction of NOR would occur. The NOR can exist in cationic form (pH < 6.2), zwitterionic/neutral form (6.2 < pH < 8.5), or anionic form (pH > 8.5) [8,12]. In acidic conditions, a large amount of H^+ ions surrounds the surface of MNPC, which could compete with the NOR molecule existing in the cationic form, and so the binding of NOR to adsorbent is restricted. When the pH value ranges from 6.0 to 8.7, the ratio of the zwitterion form is increased, so the competition between H^+ and NOR ions for surface adsorption sites is decreased correspondingly, resulting in an improved adsorption capacity. However, when the pH is higher than the pKa_2 of NOR, the anionic form dominates and the repulsion between the

NOR molecule and the negatively charged MNPC is increased, and so the adsorption capacity is significantly decreased. The further studies were conducted at the optimum pH value of 6.0.

The initial NOR concentration is another key factor controlling the adsorption performance of MNPC, as shown in Figure 9a. The initial NOR concentration ranged from 1.0 to 100 mg L^{-1} at a pH of 6.0. Obviously, the adsorption capacity is increased with the solution concentration, which was increased from 3.04 to 55.12 mg g^{-1}. A higher initial NOR concentration meant a higher concentration gradient, which led to a high driving force, and so the NOR molecules could quickly transfer to the pores of the MNPC. As shown in Figure 9b, the adsorption capacity of MNPC for NOR increased with increased temperature (20–40 °C), suggesting that the higher temperature was beneficial to the adsorption process.

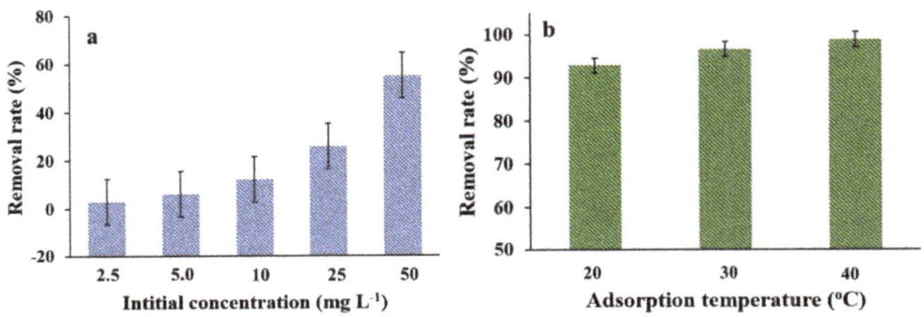

Figure 9. (**a**) The change of adsorption capacity vs initial concentration; and (**b**) the change of adsorption rates vs. temperature. All the curves were obtained in NOR solution with a dosage of 0.8 g L^{-1} at a pH of 7.0.

The influence of ionic strength on the adsorption performance was also investigated and the results are shown in Figure 10. When the salt concentration increased from 0.0 to 0.1 M, the adsorption capacity decreased slightly. Generally, the salt concentration had no significant effect on the adsorption capacity of NOR on the MNPC, which indicates that the interaction between NOR and MNPC was quite stable in a certain range of salt concentration.

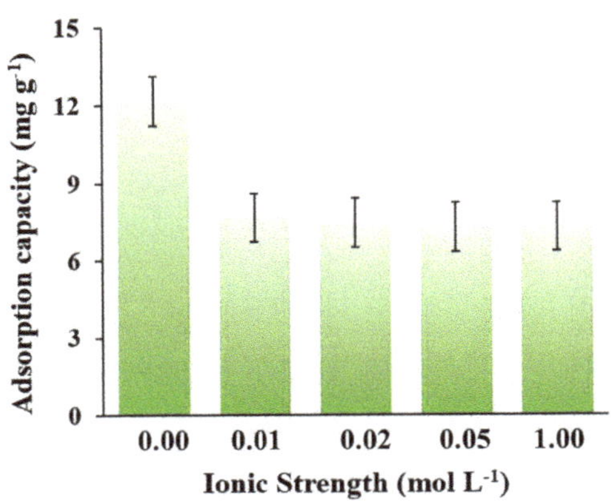

Figure 10. Plots of adsorption capacity vs ionic strength. All the curves were obtained in 10 mg L^{-1} NOR aqueous solution with a dosage of 0.8 g L^{-1} at a pH of 7.0.

3.3. Recyclability

In practical application, the recyclability is another critical factor for the adsorbents. The recyclability of the MNPC was investigated using a methanol solution (containing 10% ammonia) as the effluents and the results are shown in Figure 11. The adsorption capacity of the MNPC remained at 12.0 mg g^{-1} after five cycles, and was slightly decreased, indicating the good regeneration performance of the absorbents in the NOR solution. Hence, the MNPC adsorbent could be reused effectively, which is helpful for reducing the cost of adsorption.

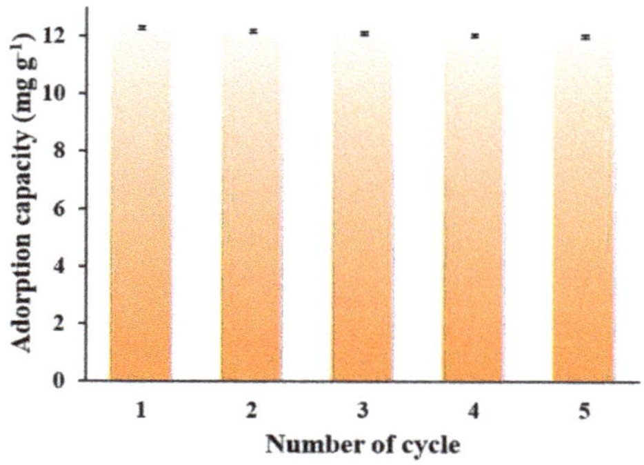

Figure 11. Regeneration property of the MNPC adsorbents in the 10 mg L^{-1} norfloxacin (NOR) aqueous solution with a dosage of 0.8 g L^{-1} at a pH of 7.

3.4. Adsorption Kinetics

Adsorption kinetic models are usually used to evaluate the variation of adsorption capacity with adsorption time, which can further reflect the relationship between adsorption performance and the structure of adsorbent. In this work, the pseudo-first-order and pseudo-second-order models were employed to analyze the experimental data. These two models are shown as follows:

$$\ln(q_e - q_t) = \ln q_e - k_1 t \tag{3}$$

$$\frac{t}{q_t} = \frac{1}{k_2 q_e^2} + \frac{t}{q_e} \tag{4}$$

where q_e and q_t (mg g^{-1}) are NOR uptakes at equilibrium, t is the adsorption time, and k_1 (min^{-1}) and k_2 (g $mg^{-1}min^{-1}$) are the rate constants of two modes [44,45].

Figure 12 shows the results of fitting the two kinetic models. As seen from the Figure 11a, the experimental data severely deviates from the fitted data, and the R^2 value of the pseudo-first order model was relatively low, indicating a low correlation of NOR adsorption kinetics data, and so the pseudo-first-order model was inconsistent with the experimental data. However, the experimental adsorption capacity values were in agreement with the theoretical adsorption capacity values according to the pseudo-second-order model (Figure 11b) with a corresponding R^2 of 0.9996, illustrating that the adsorption data fit well to the pseudo-second-order kinetic model.

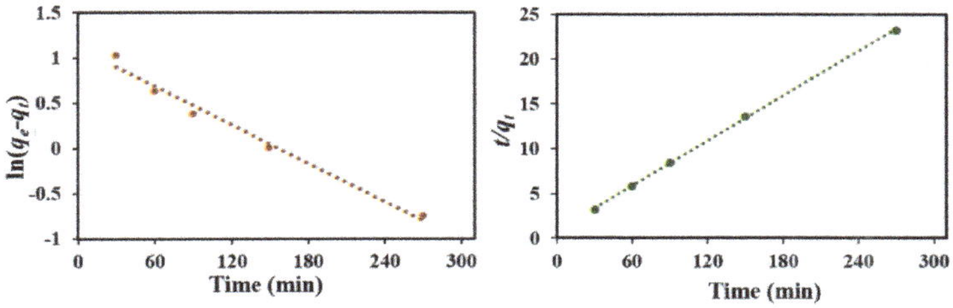

Figure 12. Adsorption kinetics of NOR on MNPC (**a**) pseudo-first-order and (**b**) pseudo-second-order models.

3.5. Adsorption Isotherm

To further study how the adsorbate interacts with the adsorbent, the adsorption models have been applied to understand the adsorption mechanism. Thus the Freundlich and Langmuir adsorption isotherm models were used according to:

$$q_{\mathrm{e}} = \frac{q_{\mathrm{m}} K_{\mathrm{L}} C_{\mathrm{e}}}{1 + K_{\mathrm{L}} C_{\mathrm{e}}} \tag{5}$$

$$q_{\mathrm{e}} = K_{\mathrm{F}} C_{\mathrm{e}}^{1/\mathrm{n}} \tag{6}$$

where q_e, q_m, and C_e (mg g^{-1}) are the adsorption capacity, equilibrium concentration, and the maximum adsorption capacity, respectively, and K_L, K_F, and n are the Langmuir and Freundlich parameters [13]. Figure 13 displays both the experimental data and the fitting isotherms of the above two isotherm models. According to the results, the adsorption isotherm of NOR onto the MNPC fits Freundlich isotherm model with higher correlation coefficients R^2 values (0.9988) compared with the Langmuir isotherm model (0.9841, Figure S6), indicating that the adsorption process predominantly features multilayer adsorption.

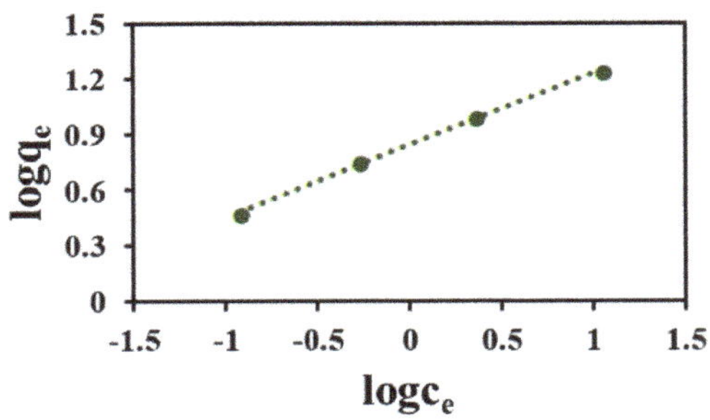

Figure 13. Freundlich isotherms for the adsorption of NOR by MNPC at 30 °C.

3.6. Adsorption Thermodynamics

The adsorption of NOR on the MNPC was further demonstrated by evaluation of changes in the Gibbs free energy ($\triangle G^{\theta}$), enthalpy ($\triangle H^{\theta}$), and entropy ($\triangle S^{\theta}$) as follows:

$$\Delta G^{\theta} = -RT\ln K_{\mathrm{c}} \tag{7}$$

$$K_{\mathrm{c}} = \frac{C_{\mathrm{A}}}{C_{\mathrm{S}}} \tag{8}$$

where R is the ideal gas constant, T represents absolute temperature (K), C_A and C_S (mg L^{-1}) are the amount of NOR adsorbed and remained in the solution at equilibrium, respectively. After making the substituting of $\triangle G^{\theta} = \triangle H^{\theta} - T\triangle S^{\theta}$ into Equation (9):

$$\ln(K_c) = -\frac{\Delta G^{\theta}}{RT} = -\frac{\Delta H^{\theta}}{RT} + \frac{\Delta S^{\theta}}{R} \quad (9)$$

The values of $\triangle H^{\theta}$ and $\triangle S^{\theta}$ were then calculated from the slope and intercept of the linear regression of ln(K_c) versus 1/T [41]. As calculated, the value of $\triangle H^{\theta}$ was 70.08 kJ·mol^{-1}, indicating that sorption of NOR on the MNPC was an exothermic process. Moreover, the value of $\triangle H^{\theta}$ was higher than 20 kJ·mol^{-1}, indicating the NOR sorption onto MNPC could be mainly attributed to chemisorption. Another important thermodynamic parameter is entropy $\triangle G^{\theta}$. As calculated, the $\triangle G^{\theta}$ value was negative, indicating that the adsorption could occur spontaneously.

4. Conclusions

The MNPC was successfully prepared by self-catalytic pyrolysis of bimetallic MOF with Zn and Co as metal ions and 2-methylimidazole as a ligand. The resultant MNPC possessed a large surface area, porous structure, good graphitization, and highly dispersed N species, simultaneously. The synergistic effect of the above characteristics offered MNPC excellent adsorption performances. The MNPC exhibited a dramatic enhancement in the adsorption to NOR compared with the MC derived from the MOF with only Co as the metal ion. The adsorption capacity was 55.12 mg g^{-1} with an initial concentration of 50 mg L^{-1} at 30 °C. The pseudo-second-order and Freundlich models were a good fit for adsorption kinetics and isotherm for NOR adsorption. In the process of NOR adsorption onto the MNPC, the π–π interaction, hydrogen bonding, and pore-filling significantly improved the adsorption capability. Overall, this material is a potential adsorbent for the NOR and is expected to be used for removal of other pollutants in waste water.

Supplementary Materials: The following are available online at http://www.mdpi.com/2079-4991/8/9/664/s1, Figure S1: XRD pattern of MC, Figure S2: Raman spectra of (**a**) MC and (**b**) PC, Figure S3: The XPS spectra of MC and MNPC, C1s and N1s of MC, Figure S4: N_2 sorption isotherm of MC, Figure S5: (**a**) SEM, (**b**) TEM, (**c**) HRTEM and mapping images of MC, Figure S6: Langmuir isotherm of NOR adsorption on the MNPC at 30 °C.

Author Contributions: Conceptualization, Hui, Wang and Jinlong, Yan; Methodology, Xi, Zhang and Yan, Wang; Validation, Xiangyun Han; Formal Analysis, Hui, Wang; Investigation, Xiangyun Han; Resources, Hui, Wang; Data Curation, Hui, Wang and Xiangyun Han; Writing-Original Draft Preparation, Hui, Wang.; Writing-Review & Editing, Jinlong, Yan; Visualization, Guixiang Quan; Supervision, Jinlong Yan; Project Administration, Hui, Wang; Funding Acquisition, Hui Wang.

Funding: This research was funded by the Natural Science Foundation of Jiangsu Province—Youth Fund Project (BK20170475) and National Natural Science Foundation of China (21677119).

Acknowledgments: The authors would like to thank C. J. Ma from Analysis and Test Center of Yancheng Institute of Technology, for help with the TEM measurements.

Conflicts of Interest: The authors declare no conflict of interest

References

1. Zhang, H.; Jia, Y.; Khanal, S.K.; Lu, H.; Fang, H.; Zhao, Q. Understanding the role of extracellular polymeric substances (EPS) on ciprofloxacin (CIP) adsorption in aerobic sludge, anaerobic sludge and sulfate-reducing bacteria (SRB) sludge systems. *Environ. Sci. Technol.* **2018**, *52*, 6476–6486. [CrossRef] [PubMed]
2. Xiao, T.; Tang, Z.; Yang, Y.; Tang, L.; Zhou, Y.; Zou, Z. In situ construction of hierarchical WO_3/g-C_3N_4 composite hollow microspheres as a Z-scheme photocatalyst for the degradation of antibiotics. *Appl. Catal. B* **2018**, *220*, 417–428. [CrossRef]
3. Wang, S.; Li, X.; Zhao, N.; Quan, X.; Chen, S.; Yu, H. Enhanced adsorption of ionizable antibiotics on activated carbon fiber under electrochemical assistance in continuous-flow modes. *Water Res.* **2018**, *134*, 162–169. [CrossRef] [PubMed]

4. Wu, H.; Li, C.; Che, H.; Hu, H.; Hu, W.; Liu, C.; Ai, J.; Dong, H. Decoration of mesoporous $Co_3 O_4$ nanospheres assembled by monocrystal nanodots on g-$C_3 N_4$ to construct Z-scheme system for improving photocatalytic performance. *Appl. Surf. Sci.* **2018**, *440*, 308–319. [CrossRef]
5. Wang, B.; Lv, X.L.; Feng, D.; Xie, L.H.; Zhang, J.; Li, M.; Xie, Y.; Li, J.R.; Zhou, H.C. Highly stable Zr(IV)-based metal-organic frameworks for the detection and removal of antibiotics and organic explosives in water. *J. Am. Chem. Soc.* **2016**, *138*, 6204–6216. [CrossRef] [PubMed]
6. Gao, J.; Lu, Y.; Zhang, X.; Chen, J.; Xu, S.; Li, X.; Li, X.; Tan, F. Elucidating the electrostatic interaction of sulfonic acid functionalized SBA-15 for ciprofloxain adsorption. *Appl. Surf. Sci.* **2015**, *349*, 224–229. [CrossRef]
7. Yan, B.; Niu, C.H.; Wang, J. Kinetics, electron-donor-acceptor interactions, and site energy distribution analyses of norfloxacin adsorption on pretreated barley straw. *Chem. Eng. J.* **2017**, *330*, 1211–1221. [CrossRef]
8. Peng, X.; Hu, F.; Zhang, T.; Qiu, F.; Dai, H. Amine-functionalized magnetic bamboo-based activated carbon adsorptive removal of ciprofloxacin and norfloxacin: A batch and fixed-bed column study. *Bioresour. Technol.* **2018**, *249*, 924–934. [CrossRef] [PubMed]
9. Jabbari, V.; Veleta, J.M.; Zarei-Chaleshtori, M.; Gardea-Torresdey, J.; Villagrán, D. Green synthesis of magnetic MOF@GO and MOF@CNT hybrid nanocomposites with high adsorption capacity towards organic pollutants. *Chem. Eng. J.* **2016**, *304*, 774–783. [CrossRef]
10. Seo, P.W.; Khan, N.A.; Jhung, S.H. Removal of nitroimidazole antibiotics from water by adsorption over metal–organic frameworks modified with urea or melamine. *Chem. Eng. J.* **2017**, *315*, 92–100. [CrossRef]
11. Lu, H.; Wang, J.; Li, F.; Huang, X.; Tian, B.; Hao, H. Highly efficient and reusable montmorillonite/Fe_3O_4/humic acid nanocomposites for simultaneous removal of Cr(VI) and aniline. *Nanomaterials* **2018**, *8*, 537. [CrossRef] [PubMed]
12. Wang, Z.Y.; Yu, X.D.; Pan, B.; Xing, B.S. Norfloxacin sorption and its thermodynamics on surface-modified carbon nanotubes. *Environ. Sci. Technol.* **2010**, *44*, 978–984. [CrossRef] [PubMed]
13. Turco, A.; Monteduro, A.G.; Mazzotta, E.; Maruccio, G.; Malitesta, C. An innovative porous nanocomposite material for the removal of phenolic compounds from aqueous solutions. *Nanomaterials* **2018**, *8*, 350. [CrossRef] [PubMed]
14. Mo, Z.W.; Zhou, H.L.; Zhou, D.D.; Lin, R.B.; Liao, P.Q.; He, C.T.; Zhang, W.X.; Chen, X.M.; Zhang, J.P. Mesoporous metal-organic frameworks with exceptionally high working capacities for adsorption heat transformation. *Adv. Mater.* **2018**, *30*, 1704350. [CrossRef] [PubMed]
15. Oveisi, M.; Asli, M.A.; Mahmoodi, N.M. MIL-Ti metal-organic frameworks (MOFs) nanomaterials as superior adsorbents: Synthesis and ultrasound-aided dye adsorption from multicomponent wastewater systems. *J. Hazard. Mater.* **2018**, *347*, 123–140. [CrossRef] [PubMed]
16. Zheng, F.; Yang, Y.; Chen, Q. High lithium anodic performance of highly nitrogen-doped porous carbon prepared from a metal-organic framework. *Nat. Comm.* **2014**, *5*, 5261. [CrossRef] [PubMed]
17. Guan, B.Y.; Yu, L.; Lou, X.W. A dual-metal-organic-frameworks derived electrocatalyst for oxygen reduction. *Energy Environ. Sci.* **2016**, *9*, 3092–3096. [CrossRef]
18. Liu, B.; Shioyama, H.; Akita, T.; Xu, Q. Metal-organic framework as a template for porous carbon synthesis. *J. Am. Chem. Soc.* **2008**, *130*, 5390–5391. [CrossRef] [PubMed]
19. Yang, S.J.; Kim, T.; Im, J.H.; Kim, Y.S.; Lee, K.; Jung, H.; Park, C.R. MOF-derived hierarchically porous carbon with exceptional porosity and hydrogen storage capacity. *Chem. Mater.* **2012**, *24*, 464–470. [CrossRef]
20. Li, S.; Zhang, X.; Huang, Y. Zeolitic imidazolate framework-8 derived nanoporous carbon as an effective and recyclable adsorbent for removal of ciprofloxacin antibiotics from water. *J. Hazard. Mater.* **2017**, *321*, 711–719. [CrossRef] [PubMed]
21. Didaskalou, C.; Buyuktiryaki, S.; Kecili, R.; Fonte, C.P.; Szekely, G. Valorisation of agricultural waste with an adsorption/nanofiltration hybrid process: from materials to sustainable process design. *Green Chem.* **2017**, *19*, 3116–3125. [CrossRef]
22. Razali, M.; Kim, J.F.; Attfield, M.; Budd, P.M.; Drioli, E.; Lee, Y.M.; Szekely, G. Sustainable wastewater treatment and recycling in membrane manufacturing. *Green Chem.* **2015**, *17*, 5196–5205. [CrossRef]
23. Likon, M.; Cernec, F.; Svegl, F.; Saarela, J.; Zimmie, T.F. Papermill industrial waste as a sustainable source for high efficiency absorbent production. *Waste Manage.* **2011**, *31*, 1350–1356. [CrossRef] [PubMed]

24. Liu, Y.; Gao, Z.; Wu, R.; Wang, Z.; Chen, X.; Chan, T.D. Magnetic porous carbon derived from a bimetallic metal-organic framework for magnetic solid-phase extraction of organochlorine pesticides from drinking and environmental water samples. *J. Chromatogr. A* **2017**, *1479*, 55–61. [CrossRef] [PubMed]
25. Jin, L.; Zhao, X.; Qian, X.; Dong, M. Nickel nanoparticles encapsulated in porous carbon and carbon nanotube hybrids from bimetallic metal-organic-frameworks for highly efficient adsorption of dyes. *J. Colloid Interface Sci.* **2018**, *509*, 245–253. [CrossRef] [PubMed]
26. Yuan, Y.; Yang, D.; Mei, G.; Hong, X.; Wu, J.; Zheng, J.; Pang, J.; Yan, Z. Preparation of konjac glucomannan-based zeolitic imidazolate framework-8 composite aerogels with high adsorptive capacity of ciprofloxacin from water. *Colloid. Surface. A* **2018**, *544*, 187–195. [CrossRef]
27. Niu, H.; Wang, Y.; Zhang, X.; Meng, Z.; Cai, Y. Easy synthesis of surface-tunable carbon-encapsulated magnetic nanoparticles: adsorbents for selective isolation and preconcentration of organic pollutants. *ACS Appl. Mater. Inter.* **2012**, *4*, 286–295. [CrossRef] [PubMed]
28. Fodi, T.; Didaskalou, C.; Kupai, J.; Balogh, G.T.; Huszthy, P.; Szekely, G. Nanofiltration-enabled in Situ solvent and reagent recycle for sustainable continuous-flow synthesis. *ChemSusChem* **2017**, *10*, 3435–3444. [CrossRef] [PubMed]
29. Schaepertoens, M.; Didaskalou, C.; Kim, J.F.; Livingston, A.G.; Szekely, G. Solvent recycle with imperfect membranes: A semi-continuous workaround for diafiltration. *J. Memb. Sci.* **2016**, *514*, 646–658. [CrossRef]
30. Li, N.; Yang, S.; Chen, J.; Gao, J.; He, H.; Sun, C. Electro-adsorption of tetracycline from aqueous solution by carbonized pomelo peel and composite with aniline. *Appl. Surf. Sci.* **2016**, *386*, 460–466. [CrossRef]
31. Chen, Y.Z.; Wang, C.; Wu, Z.Y.; Xiong, Y.; Xu, Q.; Yu, S.H.; Jiang, H.L. From bimetallic metal-organic framework to porous carbon: high surface area and multicomponent active dopants for excellent electrocatalysis. *Adv. Mater.* **2015**, *27*, 5010–5016. [CrossRef] [PubMed]
32. Torad, N.L.; Hu, M.; Ishihara, S.; Sukegawa, H.; Belik, A.A.; Imura, M.; Ariga, K.; Sakka, Y.; Yamauchi, Y. Direct synthesis of MOF-derived nanoporous carbon with magnetic Co nanoparticles toward efficient water treatment. *Small* **2014**, *10*, 2096–2107.
33. Ling, L.L.; Liu, W.J.; Zhang, S.; Jiang, H. Magnesium Oxide Embedded nitrogen self-doped biochar composites: fast and high-efficiency adsorption of heavy metals in an aqueous solution. *Environ. Sci. Technol.* **2017**, *51*, 10081–10089. [CrossRef] [PubMed]
34. Zhao, S.S.; Yan, T.T.; Wang, H.; Chen, G.R.; Huang, L.; Zhang, J.P.; Shi, L.Y.; Zhang, D.S. High capacity and high rate capability of nitrogen-doped porous hollow carbon spheres for capacitive deionization. *Appl. Surf. Sci.* **2016**, *369*, 460–469. [CrossRef]
35. Xie, Y.; Yang, W.; Wang, M.; Ge, X. Fibrous N-doped hierarchical porous carbon microspheres: Synthesis and adsorption performance. *Chem. Eng. J.* **2017**, *323*, 224–232. [CrossRef]
36. Wang, H.; Shi, L.; Yan, T.; Zhang, J.; Zhong, Q.; Zhang, D. Design of graphene-coated hollow mesoporous carbon spheres as high performance electrodes for capacitive deionization. *J. Mater. Chem. A* **2014**, *2*, 4739–4750. [CrossRef]
37. Sun, L.; Tian, C.; Li, M.; Meng, X.; Wang, L.; Wang, R.; Yin, J.; Fu, H. From coconut shell to porous graphene-like nanosheets for high-power supercapacitors. *J. Mater. Chem. A* **2013**, *1*, 6462–6470. [CrossRef]
38. Peng, H.; Ma, G.; Sun, K.; Mu, J.; Zhang, Z.; Lei, Z. Formation of carbon nanosheets via simultaneous activation and catalytic carbonization of macroporous anion-exchange resin for supercapacitors application. *ACS Appl. Mater. Interfaces* **2014**, *6*, 20795–20803. [CrossRef] [PubMed]
39. Wang, H.; Zhang, D.S.; Yan, T.T.; Wen, X.R.; Zhang, J.P.; Shi, L.Y.; Zhong, Q.D. Three-dimensional macroporous graphene architectures as high performance electrodes for capacitive deionization. *J. Mater. Chem. A* **2013**, *1*, 11778–11789. [CrossRef]
40. Peng, B.; Chen, L.; Que, C.; Yang, K.; Deng, F.; Deng, X.; Shi, G.; Xu, G.; Wu, M. Adsorption of antibiotics on graphene and biochar in aqueous solutions induced by π-π interactions. *Sci. Rep.* **2016**, *6*, 31920. [CrossRef] [PubMed]
41. Peiris, C.; Gunatilake, S.R.; Mlsna, T.E.; Mohan, D.; Vithanage, M. Biochar based removal of antibiotic sulfonamides and tetracyclines in aquatic environments: A critical review. *Bioresour. Technol.* **2017**, *246*, 150–159. [CrossRef] [PubMed]
42. Ji, L.; Chen, W.; Duan, L.; Zhu, D. Mechanisms for strong adsorption of tetracycline to carbon nanotubes: A comparative study using activated carbon and graphite as adsorbents. *Environ. Sci. Technol.* **2009**, *43*, 2322–2327. [CrossRef] [PubMed]

43. Zhu, H.; Chen, T.; Liu, J.; Li, D. Adsorption of tetracycline antibiotics from an aqueous solution onto graphene oxide/calcium alginate composite fibers. *RSC Adv.* **2018**, *8*, 2616–2621. [CrossRef]
44. Liang, C.; Zhang, X.; Feng, P.; Chai, H.; Huang, Y. ZIF-67 derived hollow cobalt sulfide as superior adsorbent for effective adsorption removal of ciprofloxacin antibiotics. *Chem. Eng. J.* **2018**, *344*, 95–104. [CrossRef]
45. Han, X.; Wang, H.; Zhang, L. Efficient removal of methyl blue using nanoporous carbon from the waste biomass. *Water Air Soil Poll.* **2018**, *229*, 26. [CrossRef]

MDPI
St. Alban-Anlage 66
4052 Basel
Switzerland
Tel. +41 61 683 77 34
Fax +41 61 302 89 18
www.mdpi.com

Nanomaterials Editorial Office
E-mail: nanomaterials@mdpi.com
www.mdpi.com/journal/nanomaterials

www.ingramcontent.com/pod-product-compliance
Lightning Source LLC
Chambersburg PA
CBHW061628210926
51850CB00009B/2473
9783039288359